Manish Kumar Singh
Priyanka Singh

Handbook on Vermicomposting

Requirements, Methods, Advantages and
Applications

Anchor Academic
Publishing

Singh, Manish Kumar, Singh, Priyanka: Handbook on Vermicomposting: Requirements, Methods, Advantages and Applications. Hamburg, Anchor Academic Publishing 2014

Buch-ISBN: 978-3-95489-276-1
PDF-eBook-ISBN: 978-3-95489-776-6
Druck/Herstellung: Anchor Academic Publishing, Hamburg, 2014

Bibliografische Information der Deutschen Nationalbibliothek:
Die Deutsche Nationalbibliothek verzeichnet diese Publikation in der Deutschen
Nationalbibliografie; detaillierte bibliografische Daten sind im Internet über
http://dnb.d-nb.de abrufbar.

Bibliographical Information of the German National Library:
The German National Library lists this publication in the German National Bibliography.
Detailed bibliographic data can be found at: http://dnb.d-nb.de

Manish Kumar Singh = Indira Gandhi Agriculture University, Raipur, Chhattisgarh, India
Priyanka Singh = Pt. Ravishankar Shukla University, Raipur, Chhattisgarh, India

Both authors have contributed equally.

© Anchor Academic Publishing, Imprint der Diplomica Verlag GmbH
Hermannstal 119k, 22119 Hamburg
http://www.diplomica-verlag.de, Hamburg 2014
Printed in Germany

CONTENTS

PREFACE

The earth is constantly being loaded with waste materials. They pose danger to the world by polluting the land, air as well as water resources. Some alternatives need to be developed to extract the best out of waste. Vermicomposting presents such an alternative. It utilizes the zero valued waste materials to yield valuable biofertilizers from them. It can be used to increase the production of crops as well as improving the structure of soil, that too without damaging the environment.

"There is a place in the heart where thoughts become wishes and wishes become dreams." We express our sincere gratitude to our beloved father Shri S. N. Singh and mother Smt. Ranju Singh, who bore the weight of sacrifice with patience, whose selfless love, affection, sacrifices and blessing made my path easier and helped us to make our dreams come true. Their blessings have always been the most vital source of inspiration and motivation in our life. Our most cordial thanks go to our younger sister Namrata who inspired us constantly and moulded us into the present position. How can we express our thanks to **"God"** because there is no word to express it. So, my lord, please realize and accept our feelings.

Dedicated

To

My

Parents

CHAPTER I

INTRODUCTION

The population of the world is exploding ! This is attributed to the advancement made by science in various fields. The entire population needs to be fed. This feed is derived from the land. The process of agricultural modernization has been an important contributing factor towards this. Modern agriculture utilizes necessary inputs of fertilizers, pesticides and labour. Production has been improved through these modern technologies. This has led to adverse environmental impacts. Some of them are enlisted below :

Overuse of natural resources: It leads to

- ➤ depletion of groundwater,
- ➤ loss of forests and wild habitats,
- ➤ decline in the capacity to absorb water,
- ➤ waterlogging and increased salinity.

Contamination of the atmosphere:

- ➤ by ammonia, nitrous oxide, methane and the products of burning or
- ➤ by the spraying of pesticides and insecticides

It leads to

- ➤ ozone depletion,
- ➤ global warming and
- ➤ atmospheric pollution

Contamination of food and fodder: by residues of pesticides and antibiotics.

Contamination of water: It is caused by pesticides, nitrates, etc. and leads to

- ➤ wildlife damage,
- ➤ disruption of ecosystems and
- ➤ possible health problems in drinking water.

Resistance to pesticides: in pests and diseases including herbicide resistance in weeds.

Loss of genetic diversity: causing the displacement of traditional varieties and breeds.

One needs to use some better alternatives to sort out this situation. Vermicomposting seems to be an excellent replacement of these chemical fertilizers. Vermicompost is an odorless and clean organic material containing adequate quantities of N, P, K and several essential micronutrients. It is eco-friendly, non-toxic, consumes low energy input for composting and is a recycled biological product. The left over organic matter are decomposed to yield the precious vermicompost.

Vermicompost is an organic manure (bio-fertilizer) produced as the vermicast by earth worm feeding on biological waste material.

Vermicomposting is a process in which worms are used to convert organic materials usually wastes into a humus-like material known as vermicompost. The main purpose is to process the material as quickly and efficiently as possible.

Vermiculture is the culture of earthworms to continually increase the number of worms in order to obtain a sustainable harvest. These worms are used

- ➢ to expand a vermicomposting operation or
- ➢ to sell it to the customers.

DEFINITION OF VERMICOMPOSTING

Vermicomposting can be defined as an aerobic non-thermophilic bio-oxidation process of organic waste decomposition which depends on earthworms.

FEATURES OF VERMICOMPOSTING

- Natural
- Free from chemicals
- Eco-friendly
- Non-toxic
- Utilizes garbage
- Low energy input
- Easy to maintain

- Little or no odour

- Rich in nutrients

- Excellent for the growth of plants

Used In :

- Farms

- Agriculture

- Gardens

HISTORY

Composting has been used by farmers and gardeners since prehistoric times to recycle wastes into products that are capable of boosting plant growth. The word composting originated from Latin words com = together and post = to bring. Decomposting or vermicomposting has been known from the very beginning. The Egyptians were one of the first cultures to recognize the soil amending properties of the earthworm. Worms have been observed by such scholars as Aristotle and Charles Darwin as organisms that decompose organic matter into rich humus or compost.

Charles Darwin, the English naturalist conducted a comprehensive study of burrowing earthworms. In 1881, he published his last book "The Formation of Vegetable Mould, Through the Action of Worms, With Observations of their Habits". This book reports the feeding behaviour of these organisms and conversion of the organic matter castings which favor plant growth.

Vermiculture was started in the 1950s in the United States for the production of fish baits. Vermicompost was produced in United States and United Kingdom from organic wastes by using earthworms in the 1980s. The publication of the "Proceedings of a Workshop on the Role of Earthworms in the Stabilization of Organic Residues" in 1981, 100 years after Darwin's study, is responsible for increasing vermicomposting within and outside of the United States. The research on vermiculture was carried out by Roy Hartenstein in the US and and Clive A. Edwards in the U.K. 1980s. Commercial vermicomposting projects have been developed in many countries such as England, France, the Netherlands, Germany, Italy, Spain, Poland, the United States, Cuba,

Mexico, the Bahamas, China, Japan, Philippines, India and elsewhere in Southeast Asia, as well as Australia, New Zealand, American Samoa, Hawaii, and many countries in South America.

The first Vermitechnology Unlimited worm farm was constructed on a five acre parcel with very large oaks and some pines scattered in. It was built to utilize the natural shade rather than clear cut and then put up artificial shade or a green house operation. This farm has a total of 3,000 linear feet of worm beds which averaged 3100 lbs. of worms being produced each month.

The production of vermicompost and vermimeal started in 1979 in Philippines. The International Symposium-Workshop on Vermi Technologies for Developing Countries was held in the Philippines in 2005.

In 1972, Mary Appelhof, Michigan biology teacher first started home vermicomposting. She is also known as the mother of modern day vermiculture. A 2009 article in the New York Times, "Urban Composting: A New Can of Worms," made many more American readers aware of the potential for even apartment dwellers to try vermicomposting.

- Vermicomposting centers are numerous in Cuba. When the Soviet Union fell, it became impossible for them to import commercial fertilizer. Vermicompost has been the largest single replacement for commercial fertilizer by Cuba. In 2004, an estimated 1 million tons of vermicompost were produced on the island.

- In India, and estimated 200,000 farmers practice vermicomposting and one network of 10,000 farmers produce 50,000 metric tons of vermicompost every month.

- Farmers in Australia and the West Coast of the U.S. are starting to use vermicompost in greater quantities, fuelling the development of vermicomposting industries there.

- Scientists at several Universities in the U.S., Canada, India, Australia, and South Africa are documenting the benefits of vermicompost, providing facts and figures that support the observations of those who have used it.

TYPES OF COMPOSTING

Composting is biological conversion of organic matter into humus-like material called compost by heterotrophic microorganisms such as bacteria, fungi, actinomycetes and protozoa. The process occurs naturally. Right organisms, feed material, moisture, aerobic conditions and nutrients are needed for microbial growth. At optimum conditions, the composting process can occur at a much faster rate.

According To Its Nature

Aerobic composting: - It stands for composting in the presence of air. Organic waste are broken down quickly and is not prone to smell. It requires high maintenance as it needs to be turned on a regular basis to keep air in the system and temperatures up. It is also likely to require accurate moisture monitoring. This type of compost is good for large volumes of compost.

Anaerobic composting: - It stands for composting in the absence of air. Anaerobic composting requires low maintenance as waste is simply throw in a pile. It is a slow process. It may take years to break down. Anaerobic composting produces awful smell. The bacteria break down the organic materials into harmful compounds like ammonia and methane.

Vermicomposting: - Composting is carried out by red worms including bacteria, fungi, insects and other bugs. The broken organic materials are utilized by the others to eat. Red worms eat the bacteria, fungi and the food waste and then deposit their castings. Oxygen and moisture are required for healthy composting. It requires medium level maintenance. One needs to feed the red worms and monitor the conditions.

According To Its Use

Industrial systems: - Industrial composting systems are popularising these days as an alternative to landfills. Untreated waste breaks down anaerobically in a landfill, producing methane gas and adds to greenhouse effect. It aims at treating biodegradable waste before it enters a landfill to harm.

Agriculture: - *W*indrow composting is used in agriculture. It is the production of compost by piling organic matter or biodegradable waste such as animal manure and crop residues, in long rows (*windrows*). This method is appropriate for producing larger volumes of compost. These rows are generally turned to improve porosity and oxygen content, mix in or remove moisture and redistribute cooler and hotter portions of the pile. Windrow composting is commonly used for farm scale composting.

Home: - Home composting is the simplest way to compost. At home, composting is generally done by using composting bins or in the form of pile composting. Other methods include trench composting and sheet composting. It is a small scale process and requires less outlay of capital and labour.

The World is Catching On

Vermicomposting is being adapted globally, especially in the warmer climates. India and Cuba are among the leaders.

- **Cuba**
  When the Soviet Union fell, it became impossible for them to import commercial fertilizer. Vermicompost has been the largest single replacement for commercial fertilizer by Cuba. In 2004, about 1 million ton of vermicompost was produced on the island.
- **India**
  About 200,000 farmers practice vermicomposting and a network of 10,000 farmers produce 50,000 metric tons of vermicompost every month.
- Farmers in Australia and the West Coast of the U.S. are also starting to use vermicompost in greater quantities.

- Scientists at several Universities in the U.S., Canada, India, Australia, and South Africa are documenting the benefits of vermicompost, providing facts and figures that support the observations of those who have used it.

6

VERMICOMPOST PRODUCTION AND ITS ECONOMICS

Mitchell and Edwards (1997) studied production *Eosinea fetida* of vermicompost from feed-lot cattle manure. Significant reductions in total mass of cattle manure were obtained by the activity of earthworms. The process yielded two products: residual vermicompost, and an increase in earthworm biomass. The most successful mode of manure application was found to be surface (vertical) application which resulted in a reduction of 30% of the initial manure (dry) mass and the production of live earthworms to 4.9% of the initial manure mass (dry weight). The increase in earthworm biomass represented extraction of 7, 18, 7 and 2% of initial total C, N, S and P respectively from the manure.

Sunitha *et al.* (1997) attempted evaluation of methods of vermicomposting under open field conditions. It was aimed to evaluate methods of vermicomposting under open field conditions. The heap system was found to be better than the pit method for biodegradation of wastes. The heap recorded higher population growth with a 20.37 - 20.86 fold increase in *Eudrilus eugeniae*.

Atiyeh *et al.* (2000) studied the effects of vermicomposts and composts on plant growth in horticultural container media and soil. The results showed that vermicomposts have the potential for improving plant growth when added to greenhouse container media or soil. However, there seem to be distinct differences between specific vermicomposts and composts in terms of their nutrient contents, the nature of their microbial communities, and their effects on plant growth.

Biradar *et al.* (2000) analysed the influence of seasons on the biomass of *Eudrilus eugeniae* and vermicompost production at the Regional Research Station, Bijapur, Karnataka, during 1995-98. The results of the study showed that the rainy season was more congenial for earthworm multiplication and vermicompost production than either winter or summer.

Giraddi (2000) studied the influence of vermicomposting methods and season on the biodegradation of organic wastes. An experiment was conducted during 1997-98 at Dharwad, to study the effect of vermicomposting and season on the biodegradation of organic wastes. The biodegradation was quite efficient during rainy and winter composting as compared to summer composting. This is indicated by higher vermicompost production and lower amounts of undegraded wastes in rainy and winter composting than in summer composting.

Jeyabal and Kuppuswamy (2001) investigated on the recycling of agricultural and agro-industrial wastes for the production of vermicompost. Its response was studied in a rice-legume (black gram) cropping system during 1994-96 in Tamil Nadu, India. The study showed that bio-digested slurry and weeds was found to be an ideal combination for vermicomposting considering the nutrient content and compost maturity period. The integrated application of vermicompost, fertilizer N and bio-fertilizers increased rice yield by 15.9% over application with fertilizer N alone.

Anonymous (2004) studied the market driven eco-enterprises for livelihood security. Based on the economic viability, these enterprises included production of a biological control agent, bio-fungicide, vermicompost, bio-fertilizers, food processing, mushroom culture, ornamental fish breeding, production of handmade paper and boards from crop wastes.

Garcia-Gil *et al.* (2000) and Bulluck *et al.* (2002) reported that compost produces significantly greater increases in soil organic carbon and some plant nutrients as compared to comparison with mineral fertilizers.

Dominguez (2004) stated that vermicompost is a stabilized, finely divided peat-like material with a low C:N ratio, high porosity and high water-holding capacity, in which most nutrients are present in forms that are readily taken up by plants.

Barik *et al.* (2005) studied the effect of different farm wastes on vermicomposting. Various crop residues such as paddy straw, vegetable waste, gliricidia leaves, rice bran, wheat bran, green gram haul, and gober gas slurry and groundnut haulm were mixed with cow dung and were used as substrates for vermicomposting. The production of vermicompost was maximum with the groundnut haulm treatment (2.5kgs). This was at par with having gliricidia leaves and green gram halum.

Costa *et al.* (2005) evaluated composting process through diary temperature monitoring of piles composed of wastes from the cotton carding industry and with 3 kinds of inoculums. The results showed that the rumen as inoculum presented high temperature values in the initial phase and low values in the final phase although the stabilization occurred at the same time. The system with aeration allowed faster material stabilization as compared to the system without aeration. The intensification of turnings in the second phase decreased the composting time and reduced the final volume by 46%. The vermicomposts showed higher nutrient content as compared to the other composts produced.

Reddy *et al.* (2006) conducted a study in Tiptur taluk of Tumkur district, Karnataka to workout economics of vermicompost use in coconut with a sample size of 40 vermicompost (VC) user farmers and 20 non-vermicompost user farmers. In general, VC users incurred lower expenditure on inputs especially on fertilizer and plant protection chemicals. The variable vermicompost, though not statistically significant, had positive association with the copra output. The application of VC to coconut farms resulted in many environmental benefits such as reduction in fertilizer use, plant protection chemicals and number of irrigations given to the crop.

Chinnappa Reddy *et al.* (2007) analyzed economics of vermicompost production and economic gains from its application to the crops like, banana, coconut, coffee, and pepper. The study was carried out in Coorg, Mysore, Hassan, Kolar, Mandya, Tumkur and Bangalore districts. The study focused on two types of vermicompost production, viz., vat method and heap method with regard to the vermicompost production.

Ramamurthy *et al.* (2007) in their study conducted near Nagpur in Maharashtra reported that vermicompost application would improve the yield of citrus by 21 %, with B: C ratio of 3.21. The adoption of vermicompost application increased from 3 per cent to 28 per cent over five years. The rate of return of vermicompost worked out to be 2.92.

Weber *et al.* (2007) reported that the results of several long-term studies have shown that the addition of compost improves soil physical properties by decreasing bulk density and increasing the soil water holding capacity. Long-term beneficial effects of composted materials are also observed in soil humic substances (due to an increase in the complexity of their molecular structure, which increases the humic/fulvic acid ratio) as well as in soil sorption properties.

Lazcano *et al.* (2008) reported that the processing of waste material through controlled bio-oxidation processes, such as composting, reduces the environmental risk by transforming the material into a safer and more stable product suitable for application to soil and also reduces the transportation costs because of the significant reduction in the water content of the raw organic matter.

Vivas *et al.* (2009) assessed the impact of composting and vermicomposting on bacterial community size and structure, and microbial functional diversity of an olive-mill waste. The aim of this study was to couple biochemical and molecular methodologies for evaluating the impact of two recycling technologies (composting and vermicomposting) on a toxic organic waste. Both the recycling technologies were effective in activating the microbial parameters of the toxic waste, the vermicomposting being the best process to produce greater bacterial diversity, greater bacterial numbers and greater functional diversity. Although several identical populations were detected in the processed and non-processed materials, each technology modified the original microbial communities of the waste in a diverse way, indicating the different roles of each one in the bacterial selection.

In addition to increasing plant growth and productivity, vermicompost may also increase the nutritional quality of some vegetable crops such as tomatoes (Gutierrez-Miceli *et al.*, 2007), spinach (Peyvast *et al.*, 2008), strawberries (Singh *et al.*, 2008), lettuce (Coria-Cayupan *et al.*, 2009) and Chinese cabbage (Wang *et al.*, 2010).

Similarly, some studies show that vermicomposting leachates or vermicompost water-extracts, used as substrate amendments or foliar sprays, also promote the growth of tomato plants (Tejada *et al.* 2008), sorghum (Gutierrez-Miceli *et al.* 2008), and strawberries (Singh *et al.* 2010).

The organic foods industry in the United States has grown dramatically in the past two decades. Organic foods constitute more than 2% of all food in the U.S. and organic sales are estimated to have increased by nearly 20% annually since 1990, reaching \$13.8 billion in 2005 (OTA, 2006). U.S. regulations require that organic foods be grown without synthetic pesticides, growth hormones, antibiotics or genetic engineering.

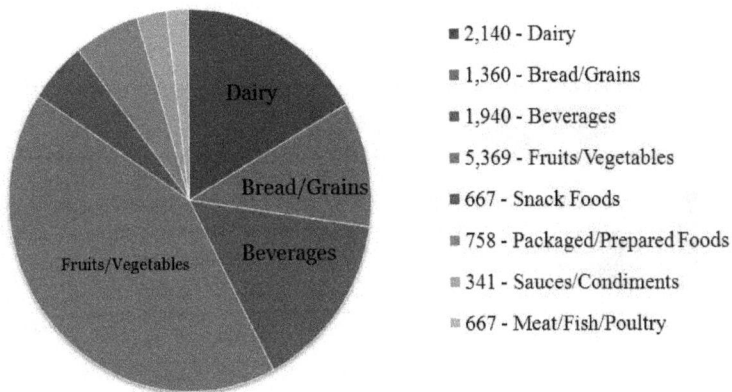

- 2,140 - Dairy
- 1,360 - Bread/Grains
- 1,940 - Beverages
- 5,369 - Fruits/Vegetables
- 667 - Snack Foods
- 758 - Packaged/Prepared Foods
- 341 - Sauces/Condiments
- 667 - Meat/Fish/Poultry

Organic food sales in the U.S. by food category, 2005, in millions of dollar, from OTA (2006).

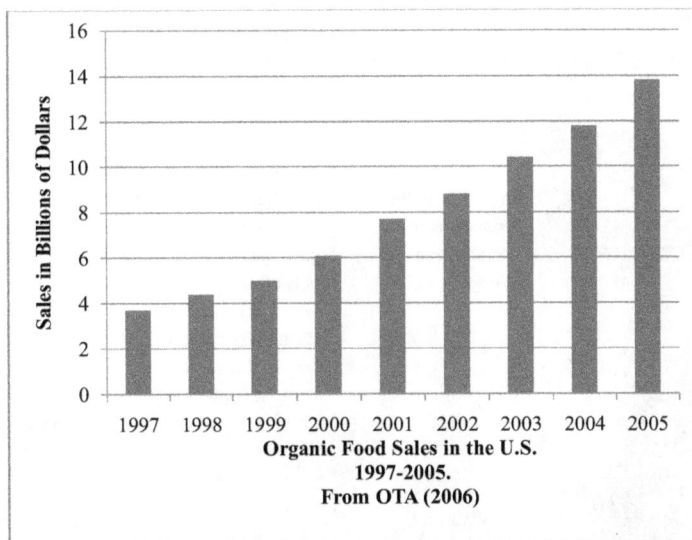

Organic Food Sales in the United States in Billions of Dollars.

Table 1: Summary of recent studies comparing organic and conventional foods with respect to nutrient levels.

Foods	Chemicals Studied	Results	Reference
Vegetable soups	Flavonols, phenolic acids	Strawberries, blueberries Organic cultivation had no consistent effects on phenolic levels	Hakkinen and Torronen (2000)
	Salicylic acid	soups had significantly higher content of salicylic acid	Baxter and others (2001)
Qing-gen-cai, Chinese cabbage, spinach, Welsh onion, green pepper	Flavonoids	Organic foods generally had higher levels of flavonoids	Ren and others (2001)
Peach, pear	Polyphenoloxidase enzyme activity, total phenolics	Organic peaches and pears had higher phenolic and polyphenol oxidase levels	Carbonaro and Mattera (2001)
Black currants	Flavonols	No consistent differences were noted between flavonol levels in organic and conventional black currants	Mikkonen and others (2001)
Peach, pear	Polyphenoloxidase enzyme activity, total phenolics, organic acids	Organic peaches and pears had higher phenolic and polyphenol oxidase levels, organic peaches had higher levels of ascorbic acid and citric acid	Carbonaro and others (2002)
Marionberries, corn, strawberries	Phenolics and ascorbic acid	Phenolics and ascorbic acid higher in organics than in conventional; highest levels of phenolics and ascorbic acid in crops grown sustainably"	Asami and others (2003)
Tomatoes	Vitamin C, carotenoids, polyphenols	Vitamin C, carotenoids, and polyphenols than conventional when results were expressed as fresh	Caris-Veyrat and others (2004)

		matter	
Grapes	Polyphenoloxidase and diphenolase enzymes	Polyphenoloxidase enzyme levels in organic and conventional grapes did not differ; diphenolase activity times higher from organic grapes than from conventional grapes	Nunez-Delicado and others (2005)
Lettuce, collards, pac choi	Phenolics	No difference in phenolic levels between organic and conventionally grown lettuce and collards; phenolics higher in organic pac choi	Young and others (2005)
Apples	Phenolics	Phenolics higher in organic apple pulp than in conventional; no differences between organic and conventional apples with respect to phenolics in apple peels	Veberic and others (2005)

Source : Organic Foods

CHAPTER II

REQUIREMENTS FOR ESTABLISHING A COMMERCIAL VERMICOMPOST UNIT

LAND

The land required to initiate a vermicompost production unit is about 0.5 – 1 acre. About 8 to 10 shacks sized 180-200 sq. ft can be built on it. A piece of land could also be taken on lease for a period of 10 to 15 years. The location of the vermicompost unit depends on various factors. Some of them are discussed below:

Objective

Small vermicomposting units are run by the farmers for meeting their own needs. It is generally observed that their homes are present in the villages and farms and fields are present away from the villages. Difficulty arises in maintenance if the vermicomposting units are established away from the home. The capacity of such units is less hence the organic waste can be brought to the home. Therefore, such units should be established near the home.

Whereas for the establishment of high capacity production units, the location should be selected on the basis of transportation facilities for raw material / wastes / manure, availability of labourer and adequate water supply.

Production capacity

The vermicompost unit should be established keeping in view the Production capacity of the unit. It can be started with any capacity and its production can be increased after a short period. The availability of worms is also an important requirement. The production of a vermicompost unit started with a capacity of 30-50 metric ton per year can be increased upto 250-300 metric ton per year. Addition of extra worms is also not required. Hence, the location of the vermicompost unit should be selected keeping in view that the production of the unit could be increased in future.

Raw material / waste / animal dung

The availability of the raw material is very important in the site selection process. Enough space should be present at the vermicompost unit site to store the raw materials for about 4 – 6 month. It is essential to maintain a stock for atleast one cycle.

Source of Water

The Source of Water should be present near the vermicompost unit. It becomes difficult to transport water from a distant source to the unit. It should also be kept in mind that the site should be at an altitude so that water logging is not found during rainy season. The vermicompost site should be well protected from the animals by building appropriate boundaries.

Shed

The vermicompost unit should be protected from direct sunlight and rain. Therefore the construction of the shed is highly recommended. The shed should be constructed on the basis of the production capacity of the unit. The hight of the unit such that no difficulty arises in carying out the various operations under 5the shed. Covering material, such as staw, cement sheet etc. can be used for the construction. The unit can be used without any boundary but it is adviced to protect the unit from sunlight. The unit should be well aerated. At least cemented walls of Four feet height should be constructed around the

Vermicompost tanks

vermicompost unit. Such types of walls are strong enough and it is easy to maintain temperature around such boundaries. It should be ascertained that the length of the shed should be from east to west.

Floor

While constructing the floor it should be kept in mind that there should not be any kind of water logging. It could be set even without flooring. But it should also be about six inches above the ground level. Such unit should be treated for termites and ants. For these chemicals such as aldosulphan dust, chloropyriphos etc. cold be used. It can also be treated herbally by using *Azadirachta indica* abstract. Bricks can be used for flooring. Before preparing the bedding on a cemented floor about 3 inches thick layer of gravel should be spread. This makes the maintenance of temperature and moisture easier.

Building

An office, warehouses for raw material & finished goods, etc. is also required.

Seed Stock

Worms are required to start vermicomposting operation. It is important to store worm seeds for future operations.

Roads, Paths & Fencing

The site should have adequate infrastructure with roads for easy movement of workers, trolleys & wheel barrows to transport the raw materials to the bed & carry out the finished compost.

Fence

Generally, vermicompost units are not attacked by animals but the site should be fenced to prevent entry of animals or unwanted elements onto the site.

Water supply, pipe line & tanks

Water is the most important ingredient after waste and worms. The vermicompost beds have to be kept moist with about 50% water content. For the production of 80 – 100 kg of vermicompost per day 20 – 30 L water is required in winter and rainy season whereas summer season 60 – 70 L water is required. It also depends upon the position of vermicomposting site, sunshine, shed, floating of the bedding, raw material etc. of the vermicomposting unit should be done men water source and pipes should be used for

water supply from distant sources. If transporting water is difficult then water can be stored bin small concrete mean the vermicompost site for smaller units and over head tanks can be used for larger units. Drippers with nonstop water flow would be viable for continuous water supply and also helps in saving water. This might be costly but reduces operational costs of manual watering. Sprinkler is the best for watering the vermicompost units. In case of pipes a sprinkler attachment at the mouth can be used.

Electricity

Electricity is essential for lighting and temperature control, such as fans to cool the worm beds and heating systems for warmth. Lights are the most effective method for preventing worms from leaving their bins and it can also be used for harvesting vermicastings as well as worms.

Machinery

For efficient operation of a vermicompost unit, the following machinery is needed.

Tubs & Trays

They are needed twice in vermicomposting operation. Once for adding the waste to the unit and secondly for taking out the finished product. But it may be required for more than two times. Difference equipments are required at the different stages of operation for e.g. for filling the bed, storage, packaging, etc. The number of tubs & trays to be used depends on the size of the unit and number of labourers employment.

Spades and shovels

These are needed to complete the vermicomposting operation Spades are used for filling the raw materials and taking out the finished products. Wooden or metallic shovels are also used. If activity of the worms is to be determined it can be done with it. Hand gloves should also be used.

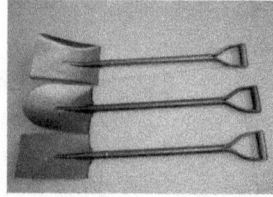

Shredder

It is used to shred or chop the hard and large sized raw material into small pieces so that it can be easily composted.

Trolleys and wheel barrows

They are used for transportation of waste organic matter, finished products, etc. to and fro the site for loading & unloading of compost.

Harvester

It is used to separate the worms from their vermicastings. It is an important process as the worms are very expensive and cannot be packed as such with the vermicastings without being separated.

Machinery for stitching & automatic packing

Vermicastings can be packed in the form of small plastic packets or sacks for smaller scale operations. Wooden boxes can be used for shipping purposes. So, machinery is required for stitching & automatic packing of vermicastings.

Transport

Transport is required for shifting raw materials to the site, especially if the source of the raw materials is far away from the unit. For a unit that produce about 1000 tones of compost per annum, a truck with the minimum capacity of 3-tonnes is required, smaller units can use smaller vehicles depending on their production. On-site vehicles, like trolleys are required to transport the raw material from the warehouses to the bedding, etc.

OPERATIONAL COSTS

The expenses include

- cost of raw materials
- fuel & Transport costs
- power
- Insurance
- repair & Maintenance
- wages for labourers
- staff salaries
- Extension Services

The number of staff & workers hired should be according to the need of workers and the size of the production unit. Manpower should be properly managed and used properly.

CHAPTER III

COMPOST WORMS

Worms are used to recycle the organic material into valuable fertilizer called vermicompost, or worm compost. The most commonly used worm for vermicomposting is the Red Wiggler (*Eisenia foetida, Eisenia fetida* or Eisenia andrei). They are also known as compost worm, manure worm and red worm. It is extremely tough and adaptable. It is found in almost all parts of the world. There are about 3,000 species of earthworms globally. Edwards & Lofty, (1972) state that the presence of about 1800 species of earthworms worldwide. The common earthworm *Lumbricus rubellus* can also be used for this purpose. The other species used for vermicomposting include:

➢ *Eisenia fetida*
➢ Eisenia veneta
➢ *Eudrilus eugeniae*
➢ *Eisenia andrei*
➢ *Eisenia hortensis*
➢ *Lumbricus terrestris*
➢ *Perionyx excavates*
➢ *Amynthas gracilis*
➢ *Lumbricus rubellus*

Desirable attributes of worms suitable for vermicomposting

1. It should have high biomass consumption rate.
2. It should also have high growth rate.
3. It should have wider tolerance range to the environmental factors.
4. It should have higher adaptability.
5. It should have higher reproductive rate.
6. It should have higher population growth rate.
7. It should show faster composting of organic residues.
8. It should have mature quickly.
9. A mixture of species is should be used.
10. It should be disease resistant.

CLASSIFICATION OF EARTHWORMS

Earthworms are often classified based on their activity and feeding type, which affects their impacts on the soil.

Compost earthworms

These are involved in composting and are mostly found in a compost bin. They prefer warm and moist environments. A ready supply of compost raw material is required. They reproduce very quickly and are usually bright red in colour and stripy.

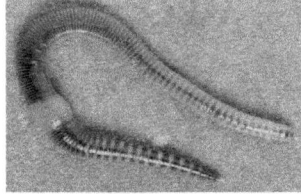

Eisenia fetida

Eg: Eisenia veneta, and *Eisenia fetida.*

Epigeic earthworms

They live on the surface of the soil in leaf litter. These species have a tendency not to make burrows but live in and feed on the leaf debris. They are also bright red or reddy-brown, but they are not stripy.

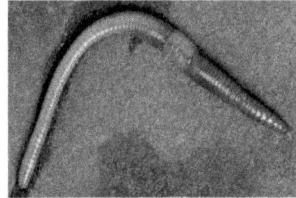

Dendrobaena octaedra

Eg: Dendrobaena octaedra, Dendrodrilus rubidus, Eiseniella tetraedra, Heliodrilus oculatus, Lumbricus rubellus, Lumbricus festivus, Lumbricus friendi, Lumbricus castaneus, Satchellius mammali, etc.

Endogeic earthworms

They live in and feed on the soil. They make horizontal burrows through the soil to move around and to feed. They reuse these burrows to a certain degree. Endogeic earthworms are often pale coloured, grey, pale pink, green or blue. Some can burrow very deeply in the soil.

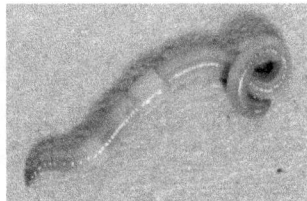

Allolobophora chlorotica

Eg : *Allolobophora chlorotica, Apporectodea*

caliginosa, Apporectodea rosea, Murchieona muldali, Octolasion cyaneum and *Octolasion tyrtaeum*, etc.

Anecic earthworms

They make permanent vertical burrows in soil. They drag leaves fallen on the soil surface into their burrows for feeding. They also cast on the surface. They also make middens (piles of casts) around the entrance to their burrows. They are darkly coloured at the head end (red or brown) and have paler tails.

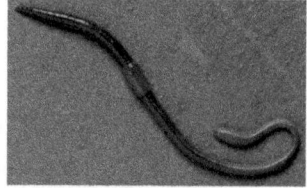

Apporectodea longa

Eg : Lumbricus terrestris, Apporectodea longa, etc.

Eisenia fetida (RED WRIGGLER)

Scientific classification	
	Kingdom : Animalia
	Phylum : Annelida
	Class : Oligochaeta
	Order : Haplotaxidae
	Suborder : Lumbricina
	Superfamily : Lumbricoidea
	Family : Lumbricidae
	Species : *Eisenia*
	Sub-species : *fetida*
Common name	Brandling or Tiger worm
Origin	North America, Europe, Middle East, Central Asia to Japan
Binomial name	*Eisenia fetida*
Appearance	Colour : Red and white strips along length of the body.
	Size : Medium (60-120 mm).

Characteristic Features	❖ They are reddish in colour, with yellowish rings.
	❖ Average length is 6 – 13 cm.
	❖ Exudes a fetid smell if handled roughly, fetida means stinky.
	❖ May be easily confused with *Eisenia veneta*.
Reproduction	❖ It has the shortest lifecycle of all earthworms.

	❖ High rates of conversion and reproduction.
	
	❖ The young worms hatch 3 weeks after the eggs are laid.
	❖ They are sexually mature within another 9 weeks.
	❖ Under perfect conditions:
	   • the worm population doubles every 3 months (4 generations a year)
	   • 500 - 600 offspring per worm per year

Cocoons of Eisenia foetida	❖ Clitellum: light-colored band – Produces cocoons ❖ Each cocoons contain ~ 4 eggs ❖ Egg incubation period = 3 weeks ❖ Number of hatchlings may be 2-9 in one cocoon.
Transformation of the organic material	❖ between half and the whole of the equivalent of its body mass a day (depending on conditions: climate, food supply) ❖ under perfect conditions: 3,500 worms (approx. 1 kg) devour 1 kg kitchen waste a day ❖ 200 - 300 worms can convert a volume of 1 m² and 20 cm depth into worm humus within 60 days ❖ of 100% source material, 15% is what remains in the form of worm compost
Use	❖ Most commonly used species for vermicomposting ❖ as fish or poultry feed ❖ In Mexico, the banana worm bread is also baked.

Eudrilus eugeniae	
Scientific classification	Kingdom : Animalia Phylum : Annelida Class : Clitellata Subclass : Oligochaeta Order : Haplotaxida Family : Eudrilidae Genus : *Eudrilus* Species : *eugeniae*
Common name	The African night crawler
Origin	Native to tropical west Africa and now widespread in warm regions
Binomial name	*Eudrilus eugeniae* Kinberg, 1867
Appearance	Clour : Uniform purple-grey sheen. The segments of the brandling worm, Eisenia fetida alternate reddish-orange and brown. The posterior segments are evenly tapered to a point and the final segment is blunt. 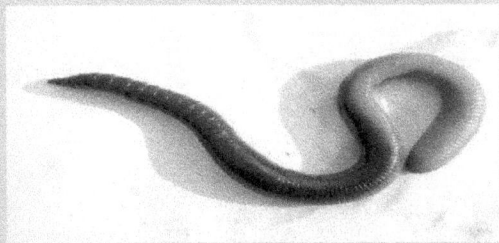
Characteristic Features	Fecundity, growth, maturation and biomass production were all significantly greater at 25°C
Use	Vermicomposting

Eisenia veneta

Scientific classification		
Kingdom	:	Animalia
Phylum	:	Annelida
Class	:	Oligochaeta
Order	:	Haplotaxidae
Suborder	:	Lumbricina
Superfamily	:	Lumbricoidea
Family	:	Lumbricidae
Species	:	*Eisenia*
Subspecies	:	*veneta*

Common name	N/A
Binomial name	*Eisenia veneta*
Origin	North America, Europe, Middle East, Central Asia to Japan

Appearance

Colour : Reddish purple, each segment has a Dark purple band, alternating with a clear intersegmental area.

Size : Medium (50-155 mm).

Characteristic Features

❖ It is epigeic
❖ Usually found in garden compost but can also occur in wet, decaying leaf litter, organic-rich soils and manure heaps

Use Vermicomposting

Eisenia hortensis (EUROPEAN NIGHTCRAWLER)

Scientific classification	Kingdom : Animalia Phylum : Annelida Class : Clitellata Order : Haplotaxida Family : Lumbricidae Genus : *Eisenia* Species : *hortensis*
Common name	European nightcrawler
Origin	European countries
Binomial name	*Eisenia hortensis* or *Dendrobaena veneta* (Michaelsen, 1890)
Appearance	Colour : Generally pink-grey in color with a banded or striped appearance. The tip of the tail is often cream or pale yellow. When the species has not been feeding, it is pale pink. Size : Medium-small earthworm averaging about 1.5 grams each when fully grown.
Characteristic Features	❖ The species is usually found in deep woodland litter and garden soils. ❖ The European nightcrawler is an invasive species that should only be used in contained compost systems and not released into wild. ❖ Compared to *E. fetida*, *E. hortensis* does best in an environment with a higher carbon to nitrogen ratio. This makes it well suited to compost pits high in fibrous materials commonly known as browns.
Use	Bait worm, but its popularity as a composting worm is increasing

Perionyx excavates

Scientific classification	Kingdom : Animalia Phylum : Annelida Class : Clitellata Subclass : Oligochaeta Order : Haplotaxida Family : Megascolecidae Genus : *Perionyx* Specie : *excavatus*
Common name	Bues or Indian blues
Origin	North America, Europe, Middle East, Central Asia to Japan.
Binomial name	*Perionyx excavatus*
Appearance	Colour : Dark brown in color with strippings. 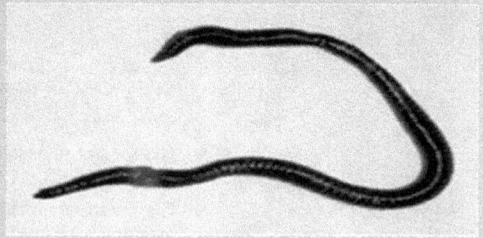
Characteristic Features	It is a commercially produced Earthworm.
Use	Good for vermicomposting in tropical and subtropical regions.

Amynthas gracilis

Scientific classification	Kingdom : Animalia Phylum : Annelida Class : Clitellata Order : Crassiclitellata Family : Megascolecidae Genus : *Amynthas* Species : *gracilis*	
Common name	*Alabama Jumpers*, Georgia Jumpers	
Origin	North America, Europe, Middle East, Central Asia to Japan.	
Binomial name	Amynthas gracilis (Kinberg, 1867)	
Appearance	Colour : Dark red in color with strippings.	
	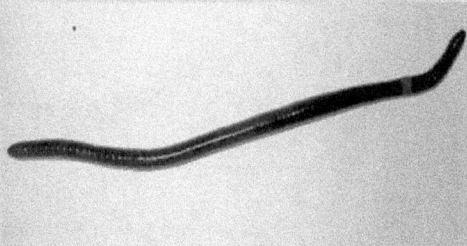	
Characteristic Features	They may be an invasive species in some areas.	
Use	Vermicomposting	

Lumbricus terrestris (COMMON EARTHWORM)

Scientific classification	Kingdom	:	Animalia
	Phylum	:	Annelida
	Class	:	Clitellata
	Subclass	:	Oligochaeta
	Order	:	Haplotaxida
	Family	:	Lumbricidae
	Genus	:	*Lumbricus*
	Species	:	*terrestris*

Common name	Britain : common earthworm or lob worm Canada : dew worm, or Grandaddy Earthworm squirrel tail, twachel or night crawler

Origin	North America, Europe, Middle East, Central Asia to Japan

Binomial name	*Lumbricus terrestris (*Linnaeus, 1758)

Appearance	Colour : Dark red or brown segmented body. Size : The largest naturally occurring species of earthworm reaching 20 - 25 cm in length when extended.

Characteristic Features	❖ *L. terrestris* is an anecic worm. ❖ It has an unusual habit of copulating on the surface at night, which makes it more visible than most other earthworms. ❖ The potential life span of *L. terrestris* is unknown ❖ *L. terrestris* is considered invasive in the north central United States.

Use	*Vermicomposting*

Lumbricus rubellus (RED EARTHWORM)

Scientific classification	Kingdom : Animalia Phylum : Annelida Class : Clitellata Subclass : Oligochaeta Order : Haplotaxida Family : Lumbricidae Genus : *Lumbricus* Species : *rubellus*
Common name	Red earthworm
Origin	Widely distributed throughout mainland Europe and the British Isles.
Distribution	All of Europe, the United States, Canada, New Zealand, and Australia. Is probably distributed worldwide, but species may not be identified in all parts of the world.
Binomial name	*Lumbricus rubellus* Hoffmeister, 1843
Appearance	Colour : Smooth, reddish, semi-transparent, flexible skin segmented into circular sections. Size : Ranges from 25-105 mm
Characteristic Features	❖ Epi-endogeic species. ❖ *Lumbricus rubellus* increases concentrations of vitamin B_{12} producing microorganisms and vitamin B_{12} in the soil. ❖ In traditional Chinese medicine, abdominal extracts from *Lumbricus rubellus* are used in a preparation known as *Di Long*, or Earth Dragon, for treatment of rheumatic, phlegm and blood disorders.
Use	❖ Vermicomposting ❖ commercial fishing bait species

BIOLOGY OF EARTHWORMS

Different species of earthworms have similar physical structure. Earthworms belong to the phylum Annelida = ringed. These rings around the worms are called segments. Redworms, *Eisenia fetida* have about 95 segments, whereas nightcrawlers, *Eisenia hortensis* have about 150 segments. They have streamlined bodies, containing no protruding appendages or sense organs, to enable them to pass easily through soil. They have well-developed nervous, circulatory, digestive, excretory, muscular, and reproductive systems.

The head or anterior end of the earthworm has a prostomium, a lobe covering the mouth that helps the earthworm in crawling in the soil. Setae (bristles) on each segment can be extended or retracted to help earthworms move. Skin glands secrete lubricating mucous which helps worms in moving through soil and stabilizes burrows and castings.

The digestive tract extends the whole length of its body. They swallow soil or residues and plant litter on the soil surface. Swallowed matter moves through the digestive tract and is mixed by strong muscles. Enzymes are secreted and blended with the materials. Simpler molecules are absorbed through intestinal membranes and are utilized by earthworms for energy and cell production. They do not have specialized organs for breathing. They breathe through their moist skin. Earthworms can live for months completely submerged in water, and die if they dry out. A red pigment in earthworms' skin makes it sensitive to ultraviolet rays. Short exposure to strong sunlight causes paralysis in some worms and longer exposure kills them. They emerge from burrows in search of oxygen when unoxygenated rainwater filters down through the soil and squeezes most of the oxygen from the soil spaces.

Earthworms are hermaphroditic, i.e., each individual possesses both male and female reproductive organs. The eggs and sperm of each earthworm are located separately to prevent self-fertilization. While mating, they face in opposite directions and exchange sperm. The eggs are fertilized at a later time. Mature eggs and sperms are deposited in a cocoon produced by the clitellum. It is a swollen, saddle-shaped structure near the worm's head. The sperm cells fertilize the eggs within the cocoon and then the cocoon slips off the worm into the soil. The number of worms inside each cocoon and the time taken to hatch varies according to worm species and environmental conditions.

Approximately *Eisenia fetida* four worms will emerge from a cocoon in 30 to 75 days, and they reach sexual maturity after 53 to 76 days.

The cocoons look a lot like grape seeds in size and shape, with one end rounded and the other slightly pointed. Cocoons are initially pearly-yellow in colour, then deepen to brown as the young inside mature and get ready to hatch. Earthworms can reproduce only by using sperm from members of their own species.

LIFECYCLE

Edwards (1988) reported the reproductive capacity of some environmentally supportive worms (Table). The lifecycle of three commonly used species for vermicomposting has been described below :

Table 2: Reproductive capacity of some environmentally supportive worms.

Species	Sexual maturity time (days)	No. of Cocoons	Cocoon hatching time (days)	Egg maturity days	Hatching (%)	No. of hatchlings	Net reproduction rate/week
E. fetida	53-76	3.8	32-73	85-149	83.2	3.3	10.4
E. eugeniae	32-95	3.6	13-27	43-122	81.0	2.3	6.7
P. excavatus	28-56	19.5	16-21	44-71	90.7	1.1	19.4
D. veneta	57-86	1.6	40-126	97-214	81.2	1.1	1.4

Source : Edwards (1988)

Eisenia foetida

It is mostly used for vermicomposting. It is also used for various toxicological studies as test worm. Mature individuals can attain up to 1.5 g body weight. Each mature worm on an average produces one cocoon every third day and from each cocoon on hatching within 23 days emerge from 1 to 3 individuals.

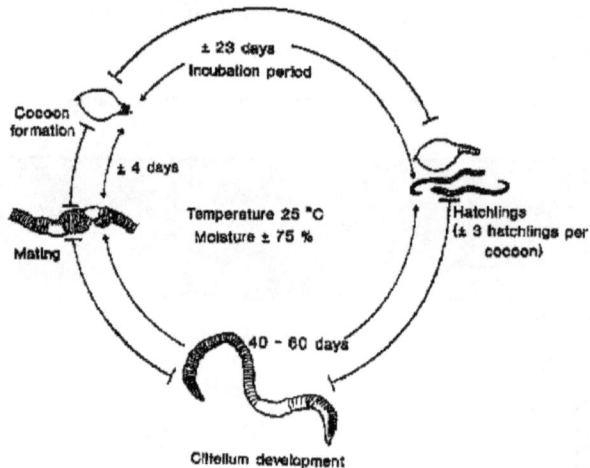

Life cycle of Eisenia foetida

Eudrilus eugeniae

It is the second most widely used earthworm for vermicomposting. It grows faster than other species. It accumulates mass at the rate of 12 mg/day. Mature individuals can attain body weight up to 4.3 g per individual. They become mature after 40 days, and, a week later, they start producing cocoon, 1 cocoon/day on average. Life span is estimated to range from 1 to 3 years. This species can be used as vermicomposting worm in tropical and sub-tropical regions.

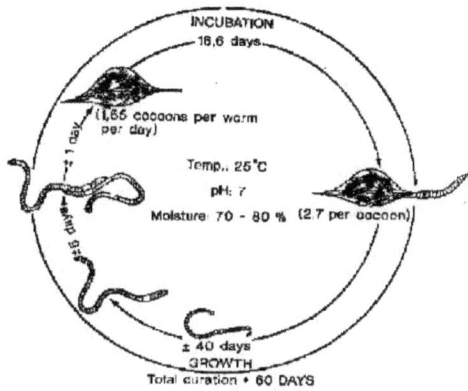

INCUBATION
16,6 days

(1,65 cocoons per worm
per day)

± 1 day

± 15 days

Temp., 25°C
pH: 7
Moisture 70 - 80 %   (2.7 per cocoon)

± 40 days
GROWTH
Total duration ± 60 DAYS

Life cycle of Eudrilus eugeniae

Perionyx excavatus

In India, this species is quite common in Eastern Himalayas, Western Himalaya, Pilibhit, Bengal and Little Andaman Islands. It shows high adaptability. It can tolerate a wide range of moisture and quality of organic matter. Average growth rate is 3.5 mg/day and body weight (maximum) 600 mg. It matures in 21-22 days and reproduces after 24 days, with 1 to 3 hatchings per cocoon.

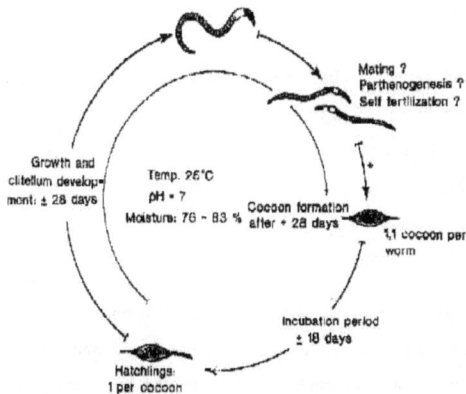

Mating ?
Parthenogenesis ?
Self fertilization ?

Growth and
clitellum develop-
ment; ± 28 days

Temp. 25°C
pH - 7
Moisture: 76 - 83 %

Cocoon formation
after ± 28 days

1,1 cocoon per
worm

Incubation period
± 18 days

Hatchlings
1 per cocoon

Life cycle of Perionyx excavates

DECOMPOSING ORGANISMS

All the micro-organisms and larger organisms involved in breaking down of organic material are called decomposing organisms. Bacteria are the primary decomposing microorganism. They break down the organic material for their own food. Bacteria grow and multiply under optimum conditions. Bacteria, actinomycetes and fungi all feed upon waste directly and are known as first level decomposers. They are assisted by larger organisms - earthworms, beetle mites, sowbugs, whiteworms, and flies. Second-level decomposers such as springtails, mold mites, feather-winged beetles, protozoa and rotifers feed upon first-level decomposing micro-organisms. Third- level decomposers such as centipedes, rove beetles, ants and predatory mites, eat both first and second-level decomposers. Organisms at each level of the food web control the population of the lower levels in check.

Bacteria

Bacteria are omnipresent. They exist on every piece of organic matter even though you can't see them. Bacteria feed upon the organic wastes and break it down into simpler forms for other bacteria and organisms to consume. Bacteria use carbon (C) as a source of energy and generate heat and carbon dioxide (CO_2) by oxidizing carbon present in the organic wastes. Nitrogen (N) is their main source of protein.

Actinomycetes

Actinomycetes are a higher form of bacteria, similar to fungi, and second in terms population after bacteria. They are especially important in the formation of humus. They liberate carbon (C), nitrate nitrogen (NO_3^-) and ammonium nitrate (NH_4^+), making nutrients available to plants.

Fungi

Fungi are lower in number as compared to bacteria or actinomycetes. But it is larger in body mass. Fungus lives on dead or dying material and obtains energy by breaking down organic materials. Fungi can also break down lignin and tannin rich organic matter.

EARTHWORM PESTS

Earthworm pests include birds, rats, snakes, moles, mice, toads and other insects or animals. Arthropods such as mites and ants are probably of the greatest concern to earthworm growers.

MITES

Mites are related to ticks, spiders, and horseshoe crabs because they have in common eight leglike, jointed appendages. They can be free-living or parasitic or both. Mites are naturally in organic materials like manure. Hence, they are also found in the compost beds. The pH of the bed should be maintained well below 6.8 to check the mites population. White or brown mites are not predaceous and compete with the worms for food. Higher mite populations can cause worms to stay deep in the beds, depriving them from the necessary nutrients resulting in poor growth and reproduction. But red mites are parasitic. It sucks the blood or body fluid of the worms and even from the egg cocoons. The adult red mite is smaller than the white or brown mite. It has bright red egg-shaped body and has eight legs.

Higher mite population denotes improperly managed beds such as

High moisture content—Too wet beds are more favourable to mites than worms. Proper drainage system and timely turning of bedding should be adopted.

Overfeeding—Too much food can cause accumulation of fermented feed in worm beds and thus lower the pH. Organic matter consumable in few days should be added at a time. Calcium carbonate can be used to adjust the pH level to neutral. Excessively wet or fleshy feed can also help the mite population to improve.

MITE REMOVAL

Proper care of worm beds can prevent the growth of harmful mites. Lower mite populations exist under ideal bed conditions.

Exposure to sunlight—The worm beds should be uncovered and exposed to sunlight for several hours. Mites may leave the worm beds.

Using newspaper—Moistened newspapers can be place on top of the beds. These papers can be removed after the accumulation of mites on them.

Using sweet fruits—Pieces of watermelon Place can be placed on top of the worm beds. Mites are attracted to the sweet fruits and accumulate on them. They can then be removed.

Heavy watering—Heavy watering (not flooding) can also help in removing mites.

Soil sulphur—Light dusting of soil sulphur can be used to kill the mites. Miticides can also be used.

ANTS

Ants are attracted to high-concentrate feed in worm beds. Some species of ants are reported to feed on eggs and small worms. Physical barriers can be placed around worm beds to keep ants out. Baits and insecticidal sprays can be used outside the bins. Pests and Diseases

MOLES

Moles naturally feed on earthworms. This is usually a problem with windrow or other open-air systems. Wire mesh, paving, or a good layer of clay can be used to prevent their attack.

BIRDS

Birds attack the worm bed only if they discover them. They will regularly come around and attack the worms. A cover of some type can be put over the material. These covers would additionally help for retaining moisture and preventing too much leaching during rainfall events.

CENTIPEDES

Centipedes have flattened and segmented body with 15 or more pairs of legs. They are third-level consumers, feeding only on insects and spiders. They can harm the worms and their cocoons. They do not multiply to a great extent within worm beds, so damage is usually low. Heavy wetting of the worm beds can be done for reducing their numbers. The water forces centipedes and other insect pests (but not the worms) to the surface, where they can be destroyed by means of a hand-held propane torch or something similar (Gaddie, op. cit.; Sherman, 1997).

SPIDERS

These are the eight-legged third-level consumers that feed on insects and small invertebrates and therefore they can harm the worms.

BEETLES

The rove beetle, ground beetle, and feather-winged beetle are the most common beetles found in compost. The rove and ground beetles prey on insects and other small animals as third-level consumers.

CHAPTER IV

RAW MATERIALS

Food Source

Regular input of feed materials is an essential step in vermicomposting process. Earthworms can feed on a wide variety of organic materials. Earthworms mainly nourish on dead and decaying organic waste and free living soil microflora and fauna. Under favourable conditions, worms can consume food higher than their body weights, Under unfavorable conditions, they can have their nourishment from soil to stay alive.

Food that is suitable to add to a worm bin includes:

- Vegetable and fruit trimmings as well as the peel
- Stale bread
- Used tea bags and tea leaves
- Toilet paper or paper towel tubes (make sure there is no glue on the roll)
- Newspapers cut into small pieces
- Vacuum cleaner dust
- Coffee grounds and filters
- Crushed eggshells (helps with worm digestion)
- Cardboard, plain and corrugated (no shiny cardboard or paperboard)
- Avocados (worms love avocados)
- Dried leaves

Food unsuitable for worm bins include:

- Citrus fruits
- Meat, including chicken or fish
- Bones
- Glossy paper (like magazines and shiny newspaper inserts)
- Salt
- Spicy vegetables (onions, hot peppers)
- Sawdust
- Dairy products
- Garden weeds

- Potato peels and sweet potato peels (could sprout in the worm bin)
- Fruit seeds
- Junk food (chips, candy, etc.)

Table 3: Common Worm Feed

Stocks Food	Advantages	Disadvantages	Notes
Cattle manure	Good nutrition; natural food, therefore little adaptation required	Weed seeds make pre-composting necessary	All manures are partially decomposed and thus ready for consumption by worms
Poultry manure	High N content results in good nutrition and a high-value product	High protein levels can be dangerous to worms, so must be used in small quantities; major adaptation required for worms not used to this feedstock. May be pre-composted but not necessary if used cautiously (see Notes)	Some books (e.g., Gaddie & Douglas, 1975) suggest that poultry manure is not suitable for worms because it is so "hot"; however, research in Nova Scotia (GEORG, 2004) has shown that worms can adapt if initial proportion of PM to bedding is 10% by volume or less.
Sheep/Goat manure	Good nutrition	Require pre-composting (weed seeds); small particle size can lead to packing, necessitating extra bulking material	With right additives to increase C:N ratio, these manures are also good beddings
Hog manure	Good nutrition; produces excellent vermicompost	Usually in liquid form, therefore must be dewatered or used with large quantities of highly absorbent bedding	Scientists at Ohio State University found that vermicompost made with hog manure outperformed all other vermicomposts, as well as commercial fertilizer

Rabbit manure	N content second only to poultry manure, there-fore good nutrition; contains very good mix of vitamins & minerals; ideal earth-worm feed (Gaddie, 1975)	Must be leached prior to use because of high urine content; can overheat if quantities too large; availability usually not good	Many U.S. rabbit growers place earthworm beds under their rabbit hutches to catch the pellets as they drop through the wire mesh cage floors.
Fresh food scraps (e.g., peels, other food prep waste, leftovers, commercial food processing wastes)	Excellent nutrition, good moisture content, possibility of revenues from waste tipping fees	Extremely variable (depending on source); high N can result in overheating; meat & high-fat wastes can create anaerobic conditions and odours, attract pests, so should NOT be included without pre-composting	Some food wastes are much better than others: coffee grounds are excellent, as they are high in N, not greasy or smelly, and are attractive to worms; alternatively, root vegetables (e.g., potato culls) resist degradation and require a long time to be consumed.

Manure

The manures obtained from cow, goat, poultry, sheep, etc. can be used for vermicomposting. It acts as a good source of nutrients but it requires pre-composting to get rid of the weed seeds.

Table 4: Livestock population and quantity of waste generated in India

Animal Species	Population (in million)	Daily average excreta animal[-1] Wet weight (kg)
Cow	185.18	11.6
Buffalo	97.92	12.2
Horse	0.75	-
Donkey	0.65	-
Sheep	61.47	0.76
Goat	124.35	0.70
Camel	0.63	-

Source : Livestock Census Report, 2003, Directorate of Economics and Statistics, Ministry of Agriculture, Government of India

Ashes

Wood ashes filtered to remove the larger particles can be of great value to composting as compared to coal ashes. It can be used against pests and contains potassium carbonate. Ash should be added in between the layers of compost waste. It is a common practice to use ash obtained by the burning of fruit skin and vegetables.

Hairs and feathers

About 6 - 7 pounds of hair contain nitrogen equivalent to 100 to 200 pounds of manure. Hair decomposes rapidly in a compost pile. But moisture and thorough mixing with an aerating material is needed. They can be bought from hair salons. Feathers of birds, chickens etc. can also be used for vermicomposting since it is a rich source of nitrogen.

Garbage

Kitchen waste, vegetable scraps are the best suites for vermicomposting. It is a rich source of nutrients and contains 1 to 3 percent nitrogen along with calcium, phosphorus, potassium, and micronutrients. Kitchen scraps can be mixed well with absorbent matter like dead leaves or hay to offset the wetness. But oily and fatty materials like meat, egg, oil, grease, etc. should be avoided. It can be covered with dirt or additional materials to discourage flies. Chopping or shredig of large pieces of matter (potatoes, grapefruit rinds, eggshells, and so on) should be done to hasten decomposition.

Potato Wastes

Potato peels, a common ingredient of kitchen scraps provides a good source of nitrogen and minor elements. Rotten potatoes can be chopped or shredded before adding to the compost pile. Potatoes contain about 2.5 percent potash. Potato vines can also be used. Dried potato contain approximately 1.6 percent potash, 4 percent calcium and 1.1 percent magnesium sulfur and other minerals.

Pea Wastes

Pea shells are rich in nitrogen and can be rapidly composted. It can be shredded or chopped before adding to the compost pile to hasten decay. Diseased pea can be burnt and the ashes can be added to the soil. Pea ash contains 3 percent phosphoric acid and 27 percent potassium.

Grasses

Grasses are rich source of organic matter and nutrients. But, it should be sun-dried to remove the excess of moisture so that it does not smell bad during composting. It can also be mixed with some dry grasses.

Stones

Stones like granite or marble also contain excess of minerals but they cannot be easily dissolved in water. Therefore, it is advisable to get them crushed into fine particles which could easily be assimilated in the compost. Similarly shells and eggshells can also be used after fine crushing.

Limestone

Limestone is an important source of calcium. It is used to raise the pH of acid soils. It should not be used with fresh manure or other nitrogenous materials due to the release of ammonia gas. It can be added directly in case of acidic soil.

Straw or Husk

Dried and weathered straw or husk can act as food for the worms. It is best to used already spoiled straw unfit for animal consumption. It is advisory to cut straw into small pieces, otherwise the worms would use a large amount of nitrogen to digest it.

Rice Hulls

Rice hulls are very rich in potash. The hulls acts an excellent soil conditioner. They can be used as mulch as well as composting. It can be obtained from rice mills. The residue left after burning rice hulls contains a high percentage of potash, making it especially valuable as a composting material.

Sawdust

It has low nitrogen content. It may cause a nitrogen deficiency if used in large amounts. But it can be used successfully as a mulch to the soil surface. It acts as a carbon source but as a bulking agent, allowing good air penetration in the pile.

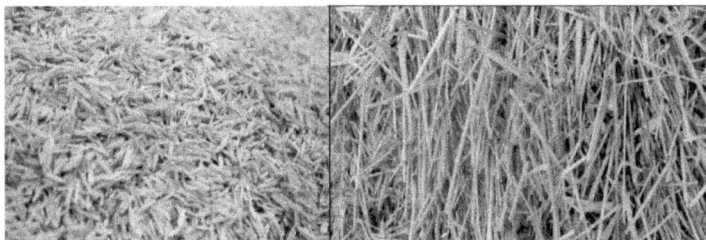

Hops

Brewery wastes can also be used for the very purpose. Hops are the plants grown and used for making beer. It is an excellent garden fertilizer. It contains 2.5 to 3.5 percent nitrogen and 1 percent phosphoric acid. But it is often found that the fresh hops are wet and have a strong odor. Their moisture content is usually high. So, additional water should not be supplied.

Molasses Residues

These are wastes obtained from sugar refining. They are rich source of carbohydrates and but they also contain some mineral nutrients. Naturally occurring yeasts in the compost will ferment these sugars rapidly. Dry molasses is also available commercially.

Leaves and twigs

Leaves carry out the process of photosynthesis. They contain a major portion of the nutrients synthesized in the plant body. Dried and rotten leaves are the best. But whole green leaves and plant twigs can also be used for vermicomposting. In such cases, the leaves and plant twigs should be chopped or shredded into small pieces. This would considerably reduce the time required for the composting process.

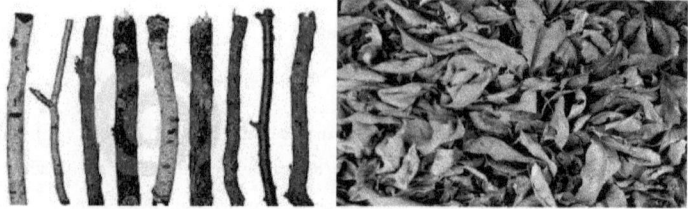

Fish Scrap and Leather dust

It can be obtained commonly near oceans or fish-processing plants. It is good source of nitrogen and phosphorus (7 percent or above for each nutrient) and also contains valuable micronutrients like iodine. Fish scraps easily turn anaerobic and attract rodents, flies etc. Fish scraps must be composted either covered with at least 4 inches of soil or in bins. Leather dust is also highly rich in nitrogen and phosphorus content.

Newspapers

They are rich source of carbon. They should be shredded into fine pieces for composting. It is generally doubted that it may be harmful to use compost prepared from newspapers as raw materials because it contains harmful inks and colours. It is debated that the ink contains polycyclic aromatic hydrocarbons which are harmful to human health and are carcinogenic. Another group of researchers argue that heavy metals are no longer used for preparing inks and natural colours are used now-a-days.

Weeds

Weeds can also be fed to the vermicomposting unit. The worms generally digest the added weed seeds as well. Therefore, there is no chance of further spread of weeds in the fields by the use of vermicompost.

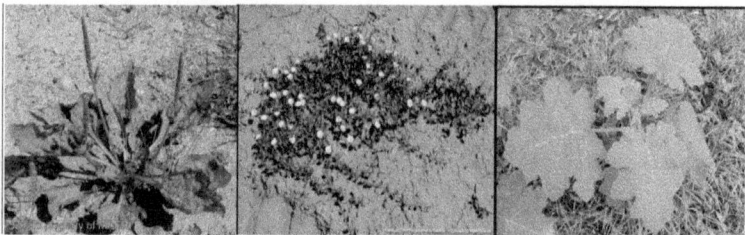

Seaweeds

They are good source of sodium, potassium, calcium, magnesium and other micronutrients. They are easy to decompose and rich source of nutrients, especially potassium. They can be mixed with straw for composting. Kelp (laminaria), bladder wrack (also called fucus), sea lettuce (ulva), and other varieties can be obtained from sea shore. It can also be bought as dried, granulated seaweed (kelp meal) or liquid concentrate. Seaweeds are rich in micronutrients. Seaweed has more potassium than manure but has less nitrogen and phosphorus. An analysis of the seaweed most commonly used in seaweed meals and extracts identified the presence of some 60 elements, including all those important for plant, animal, and human health. Fresh seaweed should be used because it decomposes rapidly. It should be mixed with nitrogenous and absorbent materials for rapid composting. Bacteria present in the alginic acid found in the leaves makes seaweed an excellent compost pile activator.

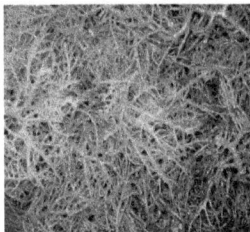

Soil

Soil is an important component in composting. The thin (1/8-inch) layer of soil used for indoor composting contains billions of soil organisms that consume plant, animal and mineral matter. These are converted to humus. It also contains minerals and organic matter. The topsoil stops heat and water from leaving the pile. Soil can be obtained from forests, fields, building excavations and streams and ponds free of industrial or agricultural pollution. Mud should be dried before composting by mixing with layers of absorbent plant wastes.

Sewage Sludge

Sewage sludge is the solid residue left after organic wastes and wastewater have been chemically, bacterially or physically processed. It may contain up to 6 percent nitrogen and from 3 to 6 percent phosphorus depending on the processing. Activated sludge is obtained by bubbling air through it. Microorganisms coagulate the organic matter and leave a clear liquid. The resulting sludge is treated thermally before using as a soil amendment followed by anaerobic fermentation for 15 to 30 days at 37°C. The dried material is incinerated, buried in landfills or as a soil conditioner. The sewage sludges are often contaminated with heavy metals due to the presence of industrial wastes. Its chemical analysis is necessary before use.

MATERIALS TO BE AVOIDED

Coal and charcoals

Coal ashes should not be added to the compost pile as they contain larger amounts of sulphur and iron, which may be harmful to the plants. Coal and charcoal generally donot decay over thousands of years. So, they should be avoided.

Diseased plants

Diseased plants should be generally avoided as it may spread diseases to other plants if the microorganisms responsible for the diseases are not destroyed during the process. For example, cabbage infected with club rot or plants suffering from leaf spot or blight should be avoided. In such cases, they should be burned to destroy the causative agents.

Nondecomposable matter

Nondecomposable matter such as rubber, plastic, metal objects, glass should never be added to the pile as they would remain there for ever and would never be decomposed.

Pet litter

It may contain disease causing agents that may affect children. For example, cat litter contains Toxoplama gondii responsible for brain and eye disease especially in children. It can also be transferred to the foetus.

Sludge

It generally requires high temperature to kill the harmful microorganisms and destroy the heavy metals. So, it should be avoided.

Toxic chemicals

Insecticides and pesticides should never be added to the vermicompost pile. They should not be applied to the area nearby an open vermicomposting site. They would kill the worms and can cause great loss. Materials treated with herbicides, insecticides and pesticides should also be avoided.

Hard materials

Husks, corn stalks, apple pumice, peanut hulls, fruit seeds etc. are hard to decompose. They should be chopped into smaller pieces for easy and faster composting.

Meat

Meat including chicken and fishes should be avoided as it is rich in fat and may yield bad smell upon decomposition. Hence, they should be avoided.

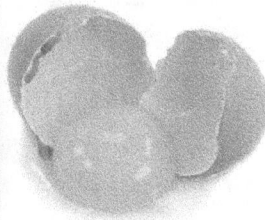

Oil and Grease

It should not be added to the compost materials as it may produce offensive smell.

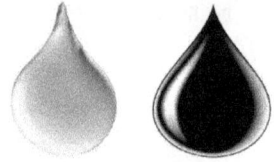

Onion

Added onion would add to the smell, so it should not be added to the bin to avoid smell.

ACCEPTABLE	CAUTIONED	UNACCEPTABLE
Fruit & vegetable	Meat and bones	Processed food
Garbage	Seeds or pits	Dairy products
Grasses	Egg shells	Oils, grease and
Hops	Citrus	Metal, plastics
Newspapers	Onion	Glass
Weeds	Ginger	Pet litter
Seaweeds	Avocado peel	Meat, fish and
Tea leaves	Hard materials	Toxic chemicals

WORM FEED

Diagram showing the suitability of various materials as worm feed.

CHAPTER V

FACTORS AFFECTING VERMICOMPOSTING PROCESS

1. Moisture

The primary requirement of worms is adequate moisture supply. It should neither be too low nor too high (may create anaerobic conditions fatal to the worms). The optimum moisture required is in the range of 60-70%.

2. Aeration

Poor aeration alongwith high levels of oily substances in the feedstock or excessive moisture may create anaerobic conditions in vermicomposting system. This may be fatal to the worms due to suffocation and toxic substances (e.g. ammonia) produced under such conditions.

3. Temperature

Temperature can greatly influence the activity, metabolism, growth, respiration and reproduction of earthworms. Most earthworm species used in vermicomposting prefer colder and moistier conditions more and require moderate temperatures from 10 – 35 ^{0}C. Tolerances and preferences may vary from species to species. Higher temperatures (> 35o C) may result in high mortality. *Eisenia* can survive having their bodies partially encased in frozen bedding and will only die when they are no longer able to consume food. Moreover, tests at the Nova Scotia Agricultural College (NSAC) have confirmed that their cocoons survive extended periods of deep freezing and remain viable (Georg, 2004).

4. Sunlight

Earthworms hate bright lights. One hour's exposure to ultraviolet rays from strong sunlight causes partial/complete paralysis and several hours are fatal. A worm breathes through its moist skin when oxygen from the air or water passes into the blood capillaries. If the body covering dries up, the worm suffocates.

5. pH

The optimum pH range for worms is 7.5 to 8.0 although they can stay alive in a pH range of 5 to 9. Generally, the pH of worm beds decreases with time due to the decomposition of organic matter under series of chemical reactions. Alkaline feed added has a moderating effect whereas neutral or acidic feed intensifies the problem. The population of pests also increases in such acidic worm beds. A solution to this problem can be the addition of calcium carbonate.

6. Surface Area:

Aerobic decomposition takes place in the compost pile by microorganisms when the particle are in direct contact with air. Surface area is directly proportional to the rate of decomposition. The surface area of the material can be increased by chopping, shredding, mowing, etc. The increased surface area leads to more digestion by microorganisms, more multiplication and more heat generation. Therefore, it is generally suggested to use the semicircular shape for piling up the organic matter for decomposition. This shape provides the largest surface area for aerobic degradation and excretion of vermicastings by the worms.

7. Worm population

The worm population in waste affects the vermicomposting process and quality. It depends on the species of worms and the type of organic waste.

8. Preparation time

organic waste materials should be composted for a while without worms before being added to the vermicompost bed. This causes the decomposition of organic matter under thermophilic condition. Thus the worms sensitive to high temperature will not damage and most of the pathogens are destroyed. Duration of the preparation affects the quality of the resulting compost, vermicomposting process and space and facilities needed for preparation.

9. Pests and Diseases

Pests such as moles and birds are a problem in outdoor vermicomposting systems. Some form of barrier, such as windrow cover, wire mesh, paving, or a good layer of clay under the windrow can be used to solve this problem. Neutral or slightly alkaline conditions should also be maintained to get rid of the blood-sucking red mites.

11. Pre-composting of organic waste

Fatty or oily feed added to the worm bin must be pre-composted, otherwise it may lead to fatal conditions for the worms under inadequency of oxygen (anaerobic conditions), as pointed out earlier.

12. Salt content

`Worms prefer salt contents less than 0.5% in feed. If manures (high soluble salt contents up to 8%) are to be used as bedding, they can be pre-composted or leached first (by running water through the material) to reduce the salt content.

12. Urine content

The manure to be used as bedding usually contain large amount of urine. It should be leached before use to remove the urine otherwise it will build up toxic gases like ammonia in the bedding.

13. Other toxic components

Different feeds can contain extensive variety of potentially toxic components. Some of the most notable are:
- De-worming medicine in manures
- Detergent cleansers industrial chemicals, pesticides.

These are usually found in sewage or septic sludge, paper-mill sludge, or some food processing wastes. Some trees, such as cedar and fir, have high levels of these naturally occurring substances. They can harm worms and even drive them from the beds. Gunadi *et al.* (2002) point out that pre-composting of wastes can reduce or even eliminate most of these threats.

- Tannins.

BARRIERS

The main barriers to optimum rates of reproduction are enlisted below:

Lack of knowledge and experience

For proper vvermicomposting one should have the knowledge and experience to get optimum results.

Lack of dedicated resources

Capital, land and labour are required for proper vermicomposting and optimum results. Lack of this may affect the process.

Lack of preparation for winter

Extreme winter conditions may reduce the rate of increase of worms even thouh they cannot completely destroy the worms. Different ways can be adopted for combating this problem.

CHAPTER VI

THE METHODS OF VERMICOMPOSTING

There are two major methods of vermicomposting, vermicomposting in bin and vermicomposting in vermicompost pile. The bin method is used for small scale operation such as home composting or garden composting. The pile method is used for vermicomposting on larger scale. It requires larger amount of wastes. It is generally applied for commercial purposes.

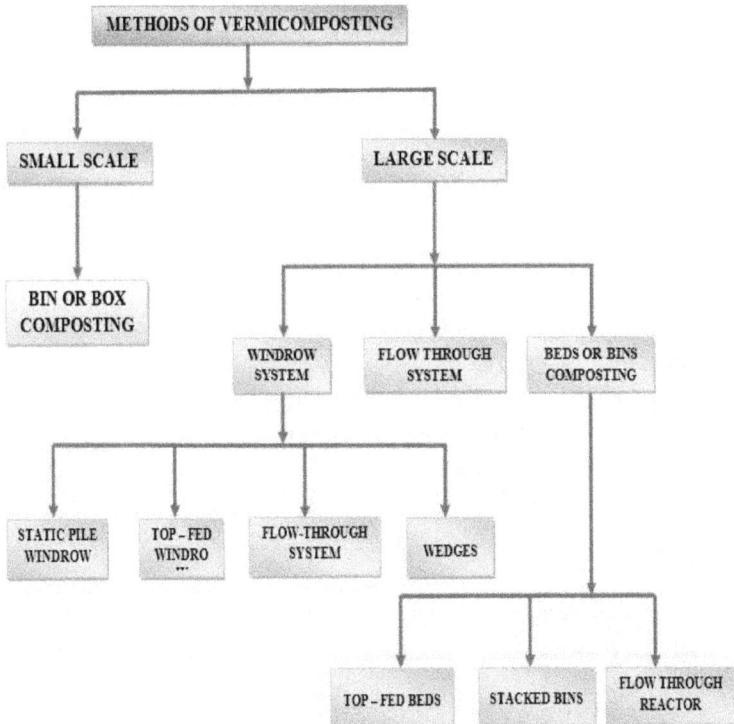

Chart showing various methods of vermicomposting.

SMALL SCALE PRODUCTION OF VERMICOMPOST

For small-scale vermicomposting at home, a large variety of bins are commercially available. Various containers such as old plastic containers, wooden or metal containers can also be used. The design of a small bin depends on the storage of the bin feeding of the worms.

Small-scale vermicomposting is best to turn kitchen waste into compost if limited space is available. Worms can decompose organic matter without the additional human physical effort. *Eisenia fetidae* are epigeic (surface dwellers) and can be used for vermicomposting but common earthworms such as *Lumbricus terrestris* are anecic(deep burrowing) species and are not suitable.

Container

Material:

Rubber

cheap, easy to use and durable.

Galvanized tubs are costlier but last forever.

Metal

often conducts heat too readily and are prone to rusting.

It may release heavy metals into the vermicompost.

Wood

Some cedars, Yellow cedar, and Redwood contain resinous oils that may harm worms, although Western Red Cedar has excellent longevity in composting conditions.

Size: It depends on

➢ the amount of organic matter to be composted,
➢ the amount of worms needed, and

> frequency of shift of storage areas.

The larger the container the greater is its capacity. Large amounts of organic matter per week will need larger areas. Longer time between bin changes will also require larger areas. The maximum productive depth for worm bin is 61cm deep.

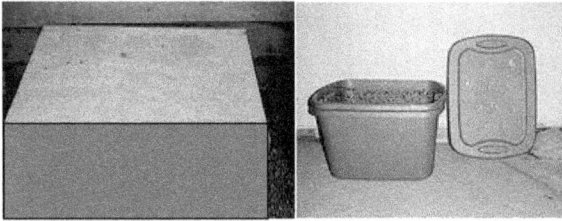

CONSTRUCTION

A worm bin is primarily the home for the worms and the place where they digest the organic material given to them. It can be purchased from the market or build at home using rubber storage totes, galvanized tubs, wood or plastic.

Ventilation: The bin should be well-ventilated, with several 3mm holes 100mm from the bottom. In case of household composting, the bin should be raised from the ground with the help of few bricks to allow drainage of fluid. For example, a worm bin can be built out of a large plastic tub with several dozen small holes on the bottom and sides.

Cover: It should have a cover for preventing light from getting in and compost from drying out.

Making shift home from old car tires.

1. A flat base having few cracks must be created from old bricks or flagstones.
2. A layer of heavy newspaper must be placed on top of the bricks.
3. The tires must be piled on top of each other.
4. The entire wormery must be filled with semi-composted organic material.
5. The composting worms must be added.
6. A piece of board weighed down with bricks can be use as a lid.
7. The worm bin should be placed in a cool area to protect it from excessive heat.
8. The outdoor temperature of the bin should be kept between 30 and 90 ^{0}Fahrenheit
9. Fertilizer can be harvested every 8 weeks (during warm months).

OPERATION

Introducing worms in the vermicomposter

The foremost requirement of a vermicomposter is to add the worms in it. The best practice is to start with a small amount of worms as these worms will feed on the organic matter and reproduce. They become larger and the population increases as well ultimately requiring more organic matter. The worms try to explore their new habitat, so they should be scrutinized closely for the first 24 hours to prevent worms from escaping the bin. The worms generally get familiarized to their new habitat within 24-48 hours no longer try to escape.

Introducing Organic matter in the vermicomposter

The organic matter should be added subsequently to the bin. Wait for the organic matter to be nearly composted before adding the new lot.

MAINTENANCE

The composter should be checked on weekly. If the worms are within the organic matter and bedding, it is an indication of an appropriately working system. If the conditions in the bedding and organic matter are not suitable for the worms, the worms would trying to escape. If no organic matter is visible and an earthy smell comes out, the vermicompost is ready to be harvested.

HARVESTING

Vermicompost can be harvested by

Handsorting – In this method vermicompost is dumped on a sheet where the worms can be picked out and put into the new composter. All worms and worm eggs (cocoons) are removed from the vermicompost. The eggs are white coloured and pea sized. They can be easily recognized and removed. A bright light can be used to concentrate the worms towards the center of the pile. This method requires more labour but is faster than other methods.

Vertical sorting – This method allows the worms to sort themselves from the vermicompost by utilizing an alternate bin. Another bin is prepared for starting a new vermicomposter. It is then placed on the top of old vermicomposter and the lid of the bin is removed. New organic feed must be added to the new vermicomposter. The starving worms start migrating towards the new food source over time. This method requires less labour but is slower than other methods.

Horizontal sorting – It is similar to vertical sorting only difference is that it does not require a second storage area. The bin is partitioned into areas and organic matter is added in a linear fashion. The worms would now migrate along the newer feed. Finally, vermicompost can be harvested at the other side of the bin. This method requires minimum work and maximum time.

LARGE SCALE/COMMERCIAL PRODUCTION OF VERMICOMPOST

The basic process of vermicomposting is same for both small scale as well as large scale production units. But in case of large scale production it is very important to maintain the appropriate environment for vermicomposting and quality of the vermicompost. A proper ratio of wastes and worms, pH, moisture, etc. should be maintained. It is practiced in Canada, Italy, Japan, Malaysia, the Philippines, and the United States of America. Mainly two main methods are used for large-scale vermiculture. Although beds and bins are most commonly used by worm growers, some choose other methods, such as windrows, the wedge system, or continuous-flow reactors.

WINDROW SYSTEM :

Windrows are linear piles on the ground containing organic matter 3 feet high. They are being used extensively both in the open and under cover. It consists of bedding materials for the earthworms to live. It consists of a concrete surface to prevent predators from attacking the worms. It acts as a large bin. Organic material is added to it. As it contains abundance of organic matter for the worms to feed on, the worms do not escape through the bin even though the windrow has no physical barriers to prevent worm escape. Organic matter is added from the front side of the bed. After mulching and continuous feeding it becomes large enough and pushes across. Back side of the bed is rich in castings. The worm castings can be removed from time to time. This side of the bed is more sensitive towards the growth of grasses and weeds.

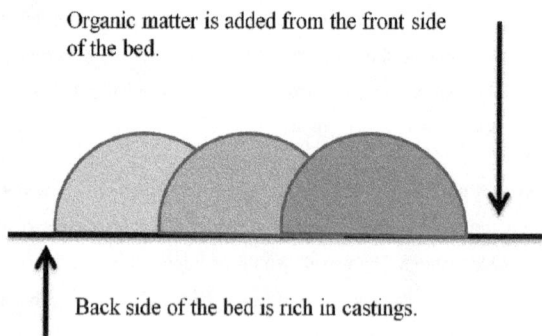

Organic matter is added from the front side of the bed.

Back side of the bed is rich in castings.

Movement of worm castings through a worm bed.

Windrow vermicomposting can be carried out in a number of different ways. The three most common are described here.

Static pile windrows (batch)

Static pile windrows are piles of mixed bedding and feed or bedding with feed layered on top. Worms and added and allowed to stand until the composting is done. These piles can be elongated, squares, rectangles or any other shape. They should not exceed one meter in height.

Nova Scotia team of researchers experimented with static windrows in 2003-04, using shredded municipally collected fibre (boxboard, cardboard, etc.) as bedding and cattle and poultry manures as feedstock.

Vermicomposting windrows of shredded cardboard and manure.

Top-fed windrows (continuous flow)

Top-fed windrows are similar to the static pile or batch windrows. The only exception is that they are not mixed and placed as a batch but are set up as a continuous-flow operation. This means that the bedding is placed first, then inoculated with worms, and then covered repeatedly with thin (less than 10 cm) layers of food. The worms consume the food at the food/bedding interface and drop their castings near the bottom of the windrow. The finished product is found on the bottom, partially consumed bedding in the middle, and the fresher food on top. Layers of new bedding should be added periodically to replace the bedding material gradually consumed by the worms.

Advantage

- ➤ The operator has greater control over the worms environment.
- ➤ The operator has a regular assess to the conditions of the windrows.

Disadvantage

> In cold countries these windrows require continuous feeding and are difficult to operate in the winter.
> If the windrows are covered, they must be removed and replaced every time the worms are fed. This creates extra work for the operator.

Wedges (continuous flow)

The wedge system is a modified windrow system. An initial stock of worms in bedding is placed inside a corral-type structure (3-sided) of no more than three feet or one meter in height. The sides of the corral can be concrete, wood, or even bales of hay or straw. The corrals can be of any width, the only limitation being contact to the interior of the piles for monitoring and remedial actions, such as adjustment of moisture content or pH level. A corral width of about 6 feet, with sufficient space between for foot travel, would be perfect. The length depends on the material being processed, worm population and other factors.

Fresh material is added on a regular feeding schedule through the open side. The worms follow the fresh food in due course, leaving the finished material behind. When the material has reached the open end of the corral, the finished product is harvested by removing the back of the corral and scooping the material out with a loader. A 4th side is then put in place and the direction is reversed.

Advantage

> The worms do not need to be separated from the vermicompost.
> During the coldest months, a layer of insulating hay or straw can be placed over the active part of the wedges.
> The regular addition of fresh manure creates enough heat to produce a temperate zone within which the worms will continue to thrive and reproduce.

FLOW-THROUGH SYSTEM :

Here the worms are fed an inch of worm chow across the top of the bed, and an inch of castings are harvested from below by pulling a breaker bar across the large mesh screen which forms the base of the bed.

Because red worms are surface dwellers constantly moving towards the new food source, the flow-through system eliminates the need to separate worms from the castings before packaging. Flow-through systems are well suited to indoor facilities, making them the preferred choice for operations in colder climates.

BEDS OR BINS

Top-fed beds (continuous flow)

It is similar to a top-fed windrow. The difference is that the bed is restricted by four walls and with/without a floor. Larger beds and bins are sheltered from the wind and precipitation. An insulating layer of straw may be used on top. Horizontal migration method is used for harvesting vermicompost.

For eg., the beds built on the Scott farm have walls of mortared cinder block and have a concrete floor inside the chicken coop, which is the lowest level of an old barn. The area receives some heat from a greenhouse attached to the building, but winter temperatures are consistently well below freezing. The bins are covered in the winter with insulating pillows made by stuffing bats of pink fibreglass insulation inside plastic bags.

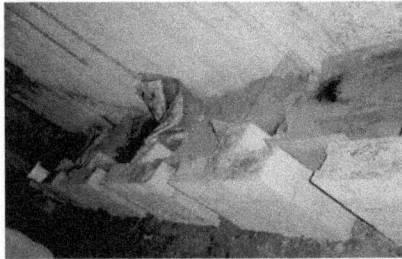

Cinder-block worm beds on Scott farm.

Stacked bins (batch or continuous flow)

Stacked bins are used when space is lacking. Vertical dimension is used to combat space related issues. The bins must be small enough to be lifted. They can be fed continuously, but this involves handling them on a regular basis. Batch process is more economical. The material is pre-mixed and placed in the bin. Then worms are added and the bin is stacked for a prefixed time period and then emptied. This method is commonly used by professional vermicomposters in North America.

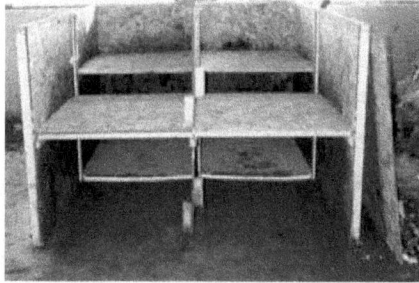

The framework for the stacked bins.

Disadvantage

➢ Initial cost of set-up is high. It requires an unheated shelter, bins, a way to mix the bedding and feed and equipments such as a forklift to stack the bins.
➢ Harvest is difficult.

Flow-Through Reactors

The flow-through concept was developed by Dr. Clive Edwards and colleagues in England in the 1980s. Since then it has been adopted and modified by several companies. The worms live in a raised box, usually rectangular and not more than three meters in width. Material is added to the top, and product is removed through a grid at the bottom, usually by means of a hydraulically driven breaker bar. The term "flow-through" refers to the fact that the worms are never disturbed in their beds – the material goes in the top, flows through the reactor and comes out at the bottom. *E. fetida* tends to eat at the surface and drop castings near the bottom of the bedding. The materials are pushed at the bottom by a set of hydraulically powered breaker bars.

SURFACE BED METHOD

It is the most commonly used method in India. The steps involved are : Windrows

1. Pre-composting

Pre-composting the feedstock decreases the amount of energy contained within the material, so that heating doesn't take occur within the worm system. Feed stock pre-composted for 10-14 days retain sufficient nutrition for the worms, but do not have much energy to generate heat. Feeds containing higher amounts of carbohydrate or woody residues should be composted beyond the heating stage. One of the advantages to the continuous flow design is in the ease with which a continuous supply of vermicompost can be removed from the system. However, harvesting of the finished material should not begin until the system is nearly full of material. Many operators have found that, along with appropriate loading rates, a minimum depth of material in the system of between 12"-18" will help to ensure that few, if any, worms will be low in the bed and drop through, or fall out with the harvested vermicompost. Once fully charged, vermicompost then
needs to be removed at a rate that maintains a relatively constant level of material in the system.

2. Preparation of Bedding

Any material that provides a fairly stable habitat to worms is called bedding. It may be constructed from lumber, concrete or cinder blocks, brick, concrete or hollow tile. Cedar, redwood or other aromatic lumber should not be used for the beds, as they contain tannic acid and resinous saps that are harmful to earthworms. Pine should also be avoided as it soaks up water and softens, allowing earthworms to eat right through it.

It should have

➢ **High absorbency:** Worms are skin-breathing. Hence the bedding must be capable to absorb and retain sufficient water.
➢ **Good bulking potential:** The bulking potential of the material should be good enough for the proper oxygen supply of worms.
➢ **High Carbon: Nitrogen ratio: Lower C:N ratio may result into** rapid degradation of the organic matter and the heat thus produced may be fatal to worms. Therefore, a higher C:N ratio is preferred.

Table 5: Common Bedding Materials

Bedding Material	Absorbency	Bulking Potential	C:N Ratio
Horse Manure	Medium-Good	Good	22 - 56
Peat Moss	Good	Medium	58
Corn Silage	Medium-Good	Medium	38 - 43
Hay – general	Poor	Medium	15 - 32
Straw – general	Poor	Medium-Good	48 - 150
Straw – oat	Poor	Medium	48 - 98
Straw – wheat	Poor	Medium-Good	100 - 150
Paper from municipal waste stream	Medium-Good	Medium	127 - 178
Newspaper	Good	Medium	170
Bark – hardwoods	Poor	Good	116 - 436
Bark -- softwoods	Poor	Good	131 - 1285
Corrugated cardboard	Good	Medium	563
Lumber mill waste -- chipped	Poor	Good	170
Paper fibre sludge	Medium-Good	Medium	250
Paper mill sludge	Good	Medium	54
Sawdust	Poor-Medium	Poor-Medium	142 - 750
Shrub trimmings	Poor	Good	53
Hardwood chips, shavings	Poor	Good	451 - 819
Softwood chips, shavings	Poor	Good	212 - 1313
Leaves (dry, loose)	Poor-Medium	Poor-Medium	40 - 80
Corn stalks	Poor	Good	60 - 73
Corn cobs	Poor-Medium	Good	56 - 123

A layer of 1.5 inch thick gravel should be spread on the surface for the preparation of vermicompost bed. This makes the maintenance of moisture in the bed easier. Gravels absorb moisture according to their capacity and let the extra water drain out. Worms and eggs also don't move down the gravel layer. Moisture moves up in case of dry atmosphere. About 3 inches thick layer of organic wastes such as crop residue, garbage, etc. should be added. The precomposted waste can be added above this layer. The waste should not contain soil.

Beds can be outdoor or indoor. Outdoor beds should be located in a shade or under open-shed roof. Indoor beds should have adequate drainage and ventilation. Bed depth should be 12 to 24 inches. If extreme temperatures (too hot or too cold) persist, beds 12 to 24 inches below ground should be constructed as the constant ground temperature will prevent the worms from freezing or overheating. The ideal distance between beds should be maintained 3 feet for carrying out various operations such as feeding, harvesting or cleaning. The lengthwise direction of the beds and their sheds should parallel the prevailing winds. For example, if the wind commonly blows from west to east, the beds should be laid out in a west-east. Worms have a tendency to creep away from their beds. Lights on over the beds all night and on rainy or foggy days can prevent the worms from moving away from the beds. Fine screens can also be fit over the beds. Moist manure, compost or old carpet placed next to the bins can also be used to capture worms that have creeped out.

Shape and size of compost waste heap

Wastes should be added in the shape of a semicircle because it provides the maximum surface area to the worms for excreting the vermicastings. Semicircular beds provide maximum aeration. Moisture also remains in the bed for longer duration (about 10-15 days). The height of the bed should be 18-20 inch and the breadth should be about 35-40 inch. The bed can be of any length depending on the production capacity needed. Sufficient distance should be maintained between two such beds for carrying out various operations.

Regular input of feed materials is an essential step in vermicomposting process. Earthworms can feed on a wide variety of organic materials. Earthworms mainly nourish on dead and decaying organic waste and free living soil microflora and fauna. Under favourable conditions, worms can consume food higher than their body weights, Under unfavorable conditions, they can have their nourishment from soil to stay alive.

Food that is suitable to add to a worm bin includes:

- Vegetable and fruit trimmings as well as the peel
- Stale bread
- Used tea bags and tea leaves
- Toilet paper or paper towel tubes (make sure there is no glue on the roll)
- Newspapers cut into small pieces
- Vacuum cleaner dust
- Coffee grounds and filters
- Crushed eggshells (helps with worm digestion)
- Cardboard, plain and corrugated (no shiny cardboard or paperboard)
- Avocados (worms love avocados)
- Dried leaves

Food unsuitable for worm bins include:

- Citrus fruits
- Meat, including chicken or fish
- Bones
- Glossy paper (like magazines and shiny newspaper inserts)
- Salt
- Spicy vegetables (onions, hot peppers)
- Sawdust
- Dairy products
- Garden weeds
- Potato peels and sweet potato peels (could sprout in the worm bin)
- Fruit seeds
- Junk food (chips, candy, etc.)

Table 6 : Common Worm Feed

Stocks Food	Advantages	Disadvantages	Notes
Cattle manure	Good nutrition; natural food, therefore little adaptation required	Weed seeds make pre-composting necessary	All manures are partially decomposed and thus ready for consumption by worms
Poultry manure	High N content results in good nutrition and a high-value product	High protein levels can be dangerous to worms, so must be used in small quantities; major adaptation required for worms not used to this feedstock. May be pre-composted but not necessary if used cautiously	Some books (e.g., Gaddie & Douglas, 1975) suggest that poultry manure is not suitable for worms because it is so "hot"; however, research in Nova Scotia (GEORG, 2004) has shown that worms can adapt if initial proportion of PM to bedding is 10% by volume or less.
Sheep/Goat manure	Good nutrition	Require pre-composting (weed seeds); small particle size can lead to packing, necessitating extra bulking material	With right additives to increase C:N ratio, these manures are also good beddings
Hog manure	Good nutrition; produces excellent vermicompost	Usually in liquid form, therefore must be dewatered or used with large quantities of highly absorbent bedding	Scientists at Ohio State University found that vermicompost made with hog manure outperformed all other vermicomposts, as well as commercial fertilizer
Rabbit manure	N content second only to poultry manure, there-fore	Must be leached prior to use because of high urine content;	Many U.S. rabbit growers place earthworm beds under

	good nutrition; contains very good mix of vitamins & minerals; ideal earth-worm feed (Gaddie, 1975)	can overheat if quantities too large; availability usually not good	their rabbit hutches to catch the pellets as they drop through the wire mesh cage floors.
Fresh food scraps (e.g., peels, other food prep waste, leftovers, commercial food processing wastes)	Excellent nutrition, good moisture content, possibility of revenues from waste tipping fees	Extremely variable (depending on source); high N can result in overheating; meat & high-fat wastes can create anaerobic conditions and odours, attract pests, so should NOT be included without pre-composting	Some food wastes are much better than others: coffee grounds are excellent, as they are high in N, not greasy or smelly, and are attractive to worms; alternatively, root vegetables (e.g., potato culls) resist degradation and require a long time to be consumed.

Water should be spread after preparing the bedding. The bed should be left as such for 2-3 days.

Preparation of vermicompost bed

3. Addition of worms

The bed temperature should be normal and it should be free of any odour before adding the worms. The worms and their eggs of about 10 inch height can be added along any one length of the bed. One advantage of this method is that worms can evaluate the favourability of the environment on their own. The worms spread all over the waste pile. Maximum population of the worms is found 2 inch below the surface and they excrete castings on the surface of the bed. About 400-500 kg of wastes and 500 worms can be added to a bed of 10 length. The ratio of waste and worms should vary in a similar ratio to that of the bed length. The waste material can be covered with a 4 inch thick finely cut straw.

4. **Removal of vermicastings**

In case of conventional composting there is no need to separate the composters from the product. The finished product can be directly put in a bag and packed. But, this is not the case with vermicomposting. The worms need to be separated from the vermicastings. The worms are too expensive to dispose of with each load of product. In batch systems such as windrows, worm harvester can be used.

After about 25-30 days about 3-4 inch thick layer vermicastings get accumulated on the bed surface. The covering of straw can be removed and the castings can be collected at the other end of the bed. Earthworms have a tendency to move away from the source of light. Hence, castings can be removed quite easily from the bedding. The waste should again be covered by the straw. It takes quite long for the preparation of first layer of vermicasting. Later on the process completes within a week. It should also be removed from the same side of the bed. At last, a small heap of waste, worms and eggs remain in the bedding.

5. **Preparation of new bed**

The wastes should be placed at one end of the bedding and again a thick layer of 4 inch can be put on the bedding. Now, the waste previously spread at the lower end of the bedding can also be fed to the worms and the straw used as covering could be used for spreading on the floor of storage room and covering the castings.

6. **Filteration of the vermicastings**

Filteration is an essential step for large scale production of vermicompost. For this a wire filter with wooden frame can be used.

7. **Packaging of vermicompost**

Vermicompost should be packed with proper amount of moisture (30-40%) and air. The packets should be kept in shady places at low temperature. It should be packed in transparent plastic packets of weight 1-5 kg. The packets should be properly sealed. Machines can be used for sealing and packaging of the vermicompost. Packets of 40-50 kg weight can be packed in H.D.P. bags. Larger amount of vermicompost can also be transported directly to the site in trucks and tractors.

Storage of vermicompost

FOUR-TANK SYSTEM

This unit is especially designed for the small farmer who approximately collects 20 - 30 kg of cattle or farm waste per day. A four-tank system, a combination of biodung composting method and vermitech technique can be set up in rural areas to enable continuous compost production. Cattle dung, weeds, leaf litter and other farm waste can be used. A tank 4m x 4m x 1m (l x b x h) is made under tree shade. This is then divided into four equal parts. The walls have vents to assist aeration as well as movement of worms from one tank to another.

The steps involved include :

1. Collection of biomass and cattle dung.
2. Biodung preparation : Soaking of biomass with water, cattle dung slurry, and covering it with black polythene sheet.
3. Collection of biomass.
4. Inoculation of earthworms.
5. Vermicompost ready and migration of earthworms from one pit to another.
6. Harvesting of compost and collection of biomass.

TWO-TANK SYSTEM

A two-tank system is suggested for household trash. For this a small shady tank of size 1m x 1m x 1m made above ground is used. It is divided into two equal halved units vertically by a wall containing vents. An average of 250 to 500 gm of trash is added daily into one of the tanks. After a few days when a layer is formed 15 to 20 cm dry/green leaves and a thin layer of soil are made to cover it. It is followed by another layering of waste over a period of time and it takes about two months for the tank to be filled. It is then covered with a black polythene sheet. The waste is now added to the second tank. After 15 to 20 days, the polythene sheet is removed, allowed to cool for a day and about 150 to 200 locally collected earthworms are released into the biomass. The biomass is converted into vermicompost within 45 to 60 days. Meanwhile, the second tank gets filled and starts decomposing. The earthworms now start migrating from the first tank to the second tank through the vents. The vermicompost is harvested from the first tank. The whole process is then repeated.

VERMICULTURE

It focuses on the production of worms, rather than vermicompost. It requires a somewhat different set of conditions than vermicomposting. The most basic differences are as follows:

Population density

A density between 5 and 10 kg/m^2 (1 to 2 lbs/ft^2) is maintained thus ensuring higher reproductive rate.

Type of system

Windrows systems have high densities only under optimum conditions. Flow-through systems are the best.

Harvesting methods

Vermiculture systems require special techniques for harvesting worms. These methods are discussed below.

Worm-growing operations can be carried out indoors as well as outdoors depending on climate, available finances and goals for worm production. The preparation of bedding and feeding of worms has already been discussed in detail.

HARVESTING

The earthworm beds should be harvested regularly. For optimize worm production it should be harvested about every 30 days. Lesser worm population provides greater organic matter to the remaining worms to feed upon and reproduce.

Bed Harvesting

Table Harvesting Technique

Beds or bins are commonly harvested by using the table harvesting technique. In this technique,

1) A table or board covered with a waterproof plastic sheet is placed next to or across the worm bed frame.
2) One or two containers with about 2 inches of pre-soaked bedding are placed next to it.
3) The top 4 inches of bedding which contain most of the worms is lifted carefully and placed on the harvesting board.
4) This operation should be carried out in the presence of brilliant sunlight or a bright light source shining overhead to drive the worms deeper into the bedding to escape the light.
5) The top inch of the bedding pile is removed gently. The worms to hideaway deeper into the bedding material.
6) The process is repeated until a solid mass of worms remains.
7) The worms are placed in the containers with the pre-soaked bedding.

Box Method

In this method a box with wooden sides and a mesh bottom is used. The steps involved are:

1) Manure or watermelon is placed in it.

2) The box is placed on the top of the worm bed.

3) The worms will move through the mesh bottom to eat the food.

4) The box can then be picked up and a new bed can be set up.

5) The worms can be packed for sale.

Another modification of the box method involves the following steps:

1) 4 square boxes 2.5 to 3 feet long are used.

2) A different size wire screen is fitted to each box with the size openings: 1/4-inch, 3/16-inch, 1/8-inch, and 1/16-inch.

3) Moist bedding and worm food is placed in each box.

4) The boxes are placed in sequence with the largest screen size on the bottom.

5) The boxes are put on the bed to be harvested or are slightly buried in it.

6) After several days the boxes are removed to find worms in each box according to size, with the largest worms trapped in the bottom box and the smallest ones in the top.

Plastic Sheet Method

In this method a plastic sheet with fresh manure is placed on top of the worm bed. Considerable numbers of worms move onto the sheet. It can be then be lifted off the bed and worms can be collected.

Windrow Harvesting

Worm growers with long 2-foot high windrows typically use a front-end loader to remove the top 6 to 8 inches of material, where most of the worms can be found. This material is deposited in a new windrow. Then the remaining castings are scooped up and screened, using a mechanical harvester.

Wedge System Harvesting

This system is basically self-harvesting because worms in the first, oldest windrow will migrate toward the fresh feed in the newer windrows. After 2 to 6 months, the first windrow and each subsequent pile can be harvested.

Continuous-Flow Reactor Harvesting

Activate a hand-operated crank or a hydraulic system to pull a bar across and just above the grill, scraping off a thin layer of finished vermicompost that falls through the widely spaced bars to the chamber or floor below.

GRADING AND COUNTING

Worms are sold by weight or by count.

There are two grades,

Bed-run   -   worms of all sizes and

Bait-size   -   worms 2.5 inches or longer when drawn up and with bodies at least 1/8 inch in diameter.

Breeding stock are large earthworms with a fully developed clitellum. Hand sorting is required for worms to be sold for bait or breeding stock. The smaller worms should be immediately put back into the beds. The larger worms should be counted or weighed. The weighing should be done in ounces.

PACKAGING AND SHIPPING

Containers specially designed for holding and shipping worms are used for the very purpose. Containers range in size from half-pints capable of holding 50 bait-size worms to gallon cartons holding 1000 bait-size or 1500-2000 bed-run earthworms. The cartons should be made of wax-coated cardboard or plastic and have small holes for air to discourage the worms from eating the containers. Moist bedding should be used to retain moisture. Canadian sphagnum peat moss might be used for this. Worms should always be stored in cool and shady areas. Clear indications like Live Earthworms, Handle With Care, Do Not Expose To Extreme Heat Or Cold, etc. should be printed on the boxes.

VERMIWASH

The vermiwash drained from the worm bed is rich in amino acids and silicic acids and is used in diluted form as foliar spray. Finished vermicompost is diluted with water to yield the liquid vermiwash.

Preparation of vermiwash

Earthworms form burrows in the soil. Bacteria, drilospheres inhabit these burrows. Water passing through these passages washes the nutrients from these burrows to the roots of the plants to be absorbed. This is the principle applied in the preparation of vermiwash. Vermiwash is a very good foliar spray. One part vermicompost is diluted with ten or twenty parts water and left to stand for about 15 to 24 hours. Then either aerobic or anaerobic methods can be followed. In aerobic method an air pump is used to pump air into the mixture throughout this period whereas in anaerobic method no air is pumped in the mixture. This method is suspected of producing substances harmful to plants due to its anaerobic nature.

(A) Vermiwash seepage tank (B) Liquid vermiwash created by rainfall is drained into collecting vessels

Application of vermiwash

The vermiwash can be either poured onto the soil or sprayed on the leaves. It strengthens the epidermis of the leaves. Thus, it reduces the damage caused by aphids and penetrating fungal spores. vermiwash can also be applied in combination with various methods of irrigation such as drip irrigation.

VERMICOMPOST TEA

The vermicompost tea is a mixture of aerobic microorganisms which extracted form vermicompost in highly aerated water. This liquid contains beneficial bacteria and fungi which help to enrich the soil, which may be poor of microorganism in result of pesticide and inorganic fertilizer application, with these microorganisms. The aerobic microorganisms also are disease-suppressive for plant. It most noted that the leachate of vermicompost during vermicomposting process is not tea it is just vermicompost leachate and may contains significant amount of not decomposed organic material.

Table 7 : Amount of vermicompost that is applied for plants

Plant	Amount of vermicompost (g)
Fruit Tree	1000-3000 According to age of the tree
Per each sapling and seedling forestry tree	100
Per each square meter of ornamental shrubs and grass	500
For ornamental plants, per square meter (flower types)	400
For each pot ,The average pots	80
For each pot, The Large pots	150

Method of tea making

A tea brewer is used for tea making. It aerates the water and extracts tea from the compost. Fresh vermicompost should be used for tea extracting as it contains more microorganisms. An incomplete vermicompost contains undecomposed organic materials that will cause quick turning the tea to anaerobic condition. It is poor in nutrients and microorganisms than a finished vermicompost. Nutrition of the microorganisms after brewing is important to keep them alive. For this purpose something such as brown sugar, honey and black strap molasses can be added to the tea.

Advantages

- Vermicompost tea has the nutrients of vermicompost.
- It is liquid and quickly reaches the plant root.
- The tea enriches soil with bacteria.
- The bacteria cover root, leaf and stalk surface and terminates anaerobic bacteria, pests and pathogens.
- It helps plants to resist against many diseases.

Limitations

It cannot be stored for a longer duration as the bacteria in the tea need food and oxygen. Food and oxygen demand of bacteria present in the tea is high. So, the bacteria will die and tea turns to anaerobic in less than a day unless the food and oxygen supply is maintained.

Tea application

- It can be used as fertilizer.
- It protects plants against diseases and pests.
- It is sprayed onto both sides of plant leaves and stalk and drenched into the root zone and used as root dip for bare root.
- It may be applied almost any time, except in cold weather conditions when soil is below 5°C.
- The UV radiation harms the microorganisms and hence it should be avoided with intense sunlight.

CHAPTER VII

PROPERTIES OF VERMICOMPOST

From thousands of years man had been been dependent solely on his nature for the fulfillment of his requirements. But with the passage of time, he began to manipulate the nature for his advantages. He introduced chemicals into the nature for his own benefit. He used insecticides, pesticides and chemical fertilizers for increasing the output from the nature. In all such attempts the nature was harmed to an irreparable limit. After all that, some group of human researchers tried to harness the gift of nature to outcome the ill-effects of these chemical monsters. But it is impossible to feed an ever-increasing population depending solely on the nature.

The canal system of irrigation was introduced in Rajasthan (India). It benefitted the people of Rajasthan to a large extent for the time being. But the ultimate effect was the destruction and conversion of thousands and lakhs of hectares of land into swamp. People use chemicals and electricity for increasing the productivity of the land. But all this is not cheap. It increases the cost of production by about 80%.

Hence Morarca Foundation carried out a research in the Jhunjhunu district of Rajasthan (India). They included 10000 farmer families of 25000 hectare area of the district. After five years of laborious research, the team concluded that the cost of production of food crops can be reduced by 80%. This can be achieved by not including the chemical fertilizers, insecticides and pesticides. They came up with idea of vermicomposting. There are lakhs of species of organisms present on the earth. Atleast 1% of them can be of benefit to the nature and hence human. Few among them can be used for vermicomposting to get rid of the chemical monsters.

PHYSICAL PROPERTIES OF VERMICOMPOST

➢ There are no fixed standards for the characterization of the prepared vermicompost.
➢ It is dark brown/black coloured soft humus-like substance. Its colour depends on the type of feed and the process of vermicomposting used. But all these things donot affect the colour of the vermicompost to much extent.
➢ It is free from faulty smell, weeds and harmful microbes.

- It consists of electrically charged particles which help the plant in abstracting nutrients from the soil.
- It is 8-10 times more nutritious than simple manure.
- It consists of a layer of mucous substance. It helps in aeration of soil and also improves the water holding capacity of the soil.
- It improves the activity of micro-organisms due to 20-30% moisture content. This improves the nutritive value of the soil.

CHEMICAL PROPERTIES OF VERMICOMPOST

On the basis of researches conducted don the compostition of vermicompost, it can be said that it contains almost all the nutrients required for all the plants. It is much better than the commercially available chemical fertilizers. It does not harm the plants in any way.

About 50% of the land in India is deficient in nitrogen. The main source of nitrogen in the soil is the organic matter solublized by the micro-organisms. These micro-organisms are killed by rise in temperature thus reducing the nitrogen content of the soil. The nitrogen content can be controlled by the addition of proper ratio of the waste materials as the feed. The digestion by worms increases the nitrogen content from 1.5-2.0 to 3.0%.

Similarly the phosphorus content is also low in the soil in India. Generally the phosphorus content is generally lower than the nitrogen content in the soil but worms fulfill this. But the phosphorus requirement is usually much higher and it can be met by the addition of external source of phosphorus.

Generally the potash content is sufficient. But some plants require higher amount of potash. At many places the waste material obtained from vegetable markets is used. It contains about 1-2% more potash.

The two most important nutrients for microbial activity and growth are nitrogen and carbon. The carbon is utilized for the growth of worms. Carbon dioxide is evolved during metabolism. It depends upon the relationship between the volatile solids content of a feedstock and CO_2 evolution. The higher the volatile solid content of organic matter, the greater is the production of carbon dioxide. Nitrogen is required for cell and protein synthesis. A high or lower carbon-to-nitrogen ratio will slow the composting process as it will release ammonia. Starches, sugars, and fats decompose or mineralize at a faster rate as compared to proteins or cellulose, whereas lignin is very resistant to mineralization.

Nath et al. (2009) reported that vermicomposting results in significant decrease in pH, Total organic carbon (TOC), electrical conductivity (EC) and C:N ratio while significant increase in total Kjeldohl nitrogen (TKN) available phosphorus, exchangeable potassium and calcium in vermicomposts/vermiwash.

Table 8 : Chemical properties of container media, vermicomposts and composts

Feature	Amount
pH	6.8 %
N_2	1 - 2.25 %
P	1 - 1.5 %
K	1.5 - 2.5 %
Ca	2.0 - 4.0 %
Na	0.02 %
Mg	0.46 %
Fe	7563 ppm
Zn	278 ppm
Mn	475 ppm
Cu	27 ppm
B	34 ppm
Al	7012 ppm
C:N	14 : 5

Source : Atiyeh et al. (2000)

BIOLOGICAL PROPERTIES OF VERMICOMPOST

Biological Degradation/Decomposition of Organics:

There are two distinct pathways of biological degradation/decomposition of organic material:

1. Anaerobic digestion

2. Aerobic digestion

Table 9 : Various properties of different types of organic wastes.

Medium	pH	Conductivity (mmhos/cm)	Total N	Total C	Total P	Total K	NH_4-N	NO_3-N
Metro-Mix 360	5.9	1.35	0.43	31.78	0.15	1.59	93	77
Vermicomposts Pig solids (VC)	5.3	4.80	2.36	43.8	4.5	0.4	123	4525
Food wastes (OS)	7.3	3.30	1.80	34.0	0.4	1.1	<14	665
Composts Biosolids (CT)	7.6	4.50	3.70	62.0	1.7	0.6	5000	165
Leaf wastes (LEAF)	8.0	1.75	1.16	35.3	0.2	0.6	32	188
Yard wastes (YARD)	8.1	1.29	0.95	33.5	0.1	0.5	24	15
Bark wastes (BARK)	7.1	0.34	0.81	70.3	0.2	0.3	18	187
Chicken Manure (CC)	6.8	2.84	4.63	51.7	3.2	3.3	4637	321

Anaerobic Digestion:

Anaerobic digestion is the break down of organic matter in the absence of oxygen under controlled conditions. Fermentation results in the formation of ammonia-like substances and hydrogen sulfide smelling like rotting eggs. Biogas containing methane, carbon dioxide and other gases is released as per following equation.

$$C_6H_{12}O_6 \longrightarrow 3CO_2 + 2CH_4 + 393kJ$$

It is more suitable for highly degradable organic waste. It requires viable anaerobic bacterial population of different species. This technology is appropriate if biogas retrieval is preferred. However, it is not recommended for composting of mixed domestic wastes. The generation of odorous gases oppose the installation of anaerobic systems in populated areas, if proper biogas recovery system is not provided.

Aerobic Digestion:

It refers to the biological degradation, vigorous humification and pasteurization of organic residues by aerobic microbes (bacteria, fungi and actinomycetes) under controlled conditions. It involves the action of mesophilic-exothermic microbes followed by thermophilic microorganisms that live at increased temperature (more than 60 ^{0}C) and if appropriately handled, can destroy pathogens. Biodegradable organic matter is decomposed releasing carbon dioxide, ammonia, water and heat. The residual organic components are converted mainly to humic acids. Finally, the desired product, organic fertilizer is obtained.

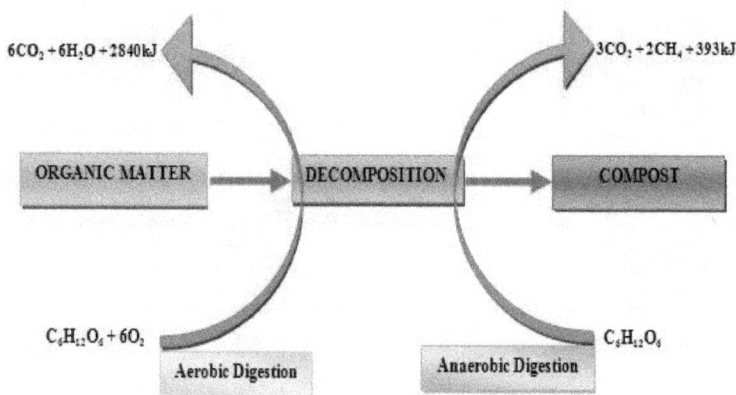

$$C_6H_{12}O_6 + 6O_2 \longrightarrow 6CO_2 + 6H_2O + 2840kJ$$

Process diagram of aerobic and anaerobic digestion of organic matter.

Vermicomposting is a complex process as compared to worms eating and excreting organic material. It is a combination of chemical, biochemical and biological reactions. This process is based on natural systems which have evolved over hundreds of millions of years. The worms eat, chew and churn up the waste. Worms play an imperative role in creating the best possible conditions for the valuable organisms to establish and reproduce. They also help in break down of wastes. The volume of waste organic matter is reduced and nutrient value is increased.

The overall mechanism involved is described below:

1. The worms ingest organic matter, fungi, protozoa, algae, nematodes and bacteria.
2. This is passed through the digestive tract.
3. The organic matter is ground into smaller particles and acted upon by the micro-organisms and enzymes present in the worms gut but majority of the bacteria and organic matter pass through undigested.
4. This forms the casting along with metabolite wastes such as ammonium, urea and proteins.
5. The worms also secrete mucus, containing polysaccharides, proteins and other nitrogenous compounds.
6. Worms create burrows in the material in this process leading to increase in the available surface area and aeration.

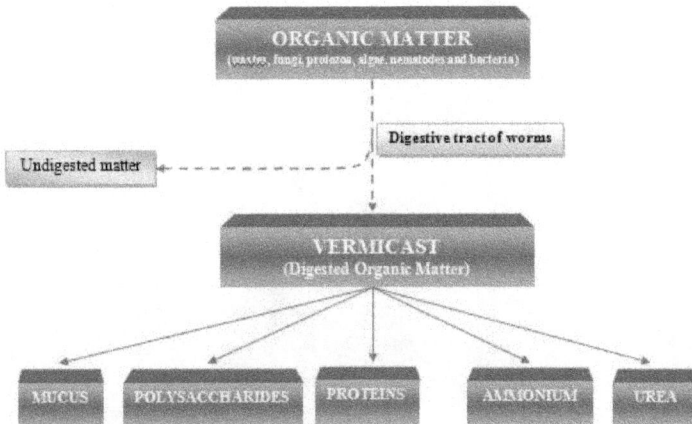

Overall mechanism of Digestion of organic wastes by worms

Some bacteria require oxygen (aerobic bacteria) whereas some object to oxygen and prefer its absence (anaerobic bacteria). Improved aeration results in favourable conditions for obligate aerobes such as Pseudomonas spp., *Zoogloea spp.*, Micrococcus spp. and *Achromobacter spp.* Anaerobic bacteria are responsible for the stench from stagnant drains, refuse sacks and landfill sites. Aerobic microbiological growth increases under aerobic conditions. It consists of nitrogen fixing bacteria, nitrification bacteria aerobic bacteria. The vermicasts have higher concentrations of soluble salts and greater nitrifying power. This microbiological growth leads to the breakdown of organic nitrogen compounds to ammonia and ammonium. The sweet smelling aerobic process overcomes the bad smell of anaerobic process. That is why properly fed and maintained worm compost piles smells good.

During initial degradation of waste of 1 m size the temperature rises to more than 70 ^{0}C in about 10 days. Hence, it can be presumed that almost all the harmful micro-organisms are destroyed during the process of vermicomposting.

CHAPTER VIII

APPLICATION OF VERMICOMPOST

VERMICOMPOST IN DRYLAND AGRICULTURE

It can be used for crops horticultural plants, road side plantation, agroforestry, pasture development, nursery etc. The hectorage of dryland agriculture is higher than irrigated agriculture in India. It depends natural rainfall and includes coarse grains like maize, millets, etc. Previously, agriculture was done on the basis of capability of the land but now-a-days chemical fertilizers, pesticides, herbicides etc. are used to increase the production and ultimately, profit. As already mentioned, these chemical monsters affect the land adversely resulting into its conversion to barren land. Various researches carried out in this area conclude that

➢ It increases the crop production always.

➢ It is best to use it before sowing.

➢ It can also be used during sprouting stage.

The benefits gained include

➢ Higher germination, more sprouting, higher crop growth, increase number, size, shape, weight etc.

➢ Some researchers have found that the effect of termides is reduced and the weed growth decreases.

➢ Food crops thus produced have higher nutrient content.

➢ Less water supply demanded.

➢ It can be used in combination manure an chemical fertilizers as well.

➢ Generally it has a threshold value above which it does have any additional effect.

➢ The crops produced with this technology also fetch higher price in the market.

VERMICOMPOST IN IRRIGATED LAND

In modern agriculture high productivity is obtained mainly because of three factors – hybrid seed, chemical fertilizers and irrigation facilities. Chemical fertilizers are used to increase the yield ignoring the capacity of land and its effect on the environment. This leads to the destruction of the soil structure. These chemical fertilizers are added in large amounts in the fields. But only 15-30% of it can be assimilated by the plants. Rest of it goes into the atmosphere and pollutes it. The run off water from the fields take the chemical fertilizers along with it to the local water bodies such as ponds, rivers, etc. and thus pollute the water resources as well. Chemical fertilizer treated crops and vegetables also affect the health of animals as well as human.

Using vermicompost improves the quantity as well as quality of the field crops.

VERMICOMPOST IN HORTICULTURE

Vermicomposting can also be used in horticultural fields and nurseries. The quality of the vegetables improves with the use of vermicompost. These vegetables gain greater attention and price in the market because of its natural size, shape and glow. In many cases, it also decreases the amount and usage of pesticides and insecticides in the field.

Table 10 : Application of Vermicompost for different fields

Banana	1 metric ton / acre
Flowers	2 metric ton / acre
Grapes	1.5 metric ton / acre
Tea	1.5 metric ton / acre
Coffee	1 metric ton / acre
Mulberry	1 metric ton / acre

VERMICOMPOST IN GARDENS AND TREE PLANTATION

Mostly people living in cities prefer gardening as a hobby because of its aesthetic value. They also produce vermicompost on a small scale from the kitchen garbage, lawn grasses, etc. This is the best way to get natural results. It can also be used in pasture development, tree plantation and agroforestry.

Table 11 : Dosage of Compost

FIELD CROPS	DOSAGE
HORTICULTURE CROPS	200 gm / plant (Young) 5 Kg / tree (Matured)
FOREST	200 gm / plant (Young)
ORNAMENTAL	50 gm / pot

VERMICOMPOST IN SOLID WASTE MANAGEMENT

It is an important required for successful vermicomposting. The organic (biological) matter is known as biomass. The following there categories of disposal methods of biomass are adopted in India.

Physical Methods: In this method, the waste material, such as crop & plant wastes, waste from cities and animal dung are first dried and reduced in volume. This reduces the weight or volume of the wastes. Faster results can be obtained by cropping the wastes. In recent years, bracketing and polytisation are in common practice.

Chemical Methods

These are of 2 types. It includes the high temperature processes such as combustion, gasification, pyrolysis, direct liquification, etc. In all these methos waste materials are transformed for energy production. The second method follows the biochemical pathway. It includes the digestion and fermentation constitutes the examples.

Combined Methods

Depending on the nature and amount of wastes, a combination of methods is used. Wastes generated from big cities, metro cities, industrial wastes, etc. are treated by these methods.

Previously, the wastes are disposed by burning. But it generates CO_2 which is a major air pollutant and are of the main culprit of global warming. Here it is not much appreciated and has been abandoned. In small cities the solid wastes are dumped in the landfills. The slums in Delhi are located on such dumping grounds. But this may result into the leaking of toxic chemicals and may pollute the ground waste. Such places become inappropriate for the point of view of living and agriculture.

In various parts of India garbage, crop residues and animal dung are mixed and left as mulch. This results into increase in the population of harmful pathogens like flies and mosquitoes which result into the spread of diseases. These wastes are then used after some time as the local manure for crops. This type of improperly decomposed solid waste also affects the crops adversely.

All these wastes, especially village wastes can be utilized for the composting and vermicomposting operations. It proves to be the best method for the disposal of solid wastes. In the Alvar district of Rajasthan, a major project is ran from the year 2000 in about 1000 villages. The Supreme Court of India has made a decision to essentially use the cit organic waste for compost production. Due to this the Central and State Government provide various legal exemptions and subsidies and even a relaxation in income tax up to an income of about Rs. 5 lakh from the composting of such wastes.

Organic Method

The epigenic worms used in vermicomposting use only organic material as food. Organic materials mainly include crop residues, animal dung and garbage. It is customary to know the actual composition of the organic material for vermicomposting. The C : N ratio can also be found by chemical analysis. Nitrogen, phosphorus, potash content and amount of micronutrients such as calcium, magnesium, zinc, iron etc can also be found. It is essential to know their content so that an effective combination of all these minerals could be provided to the plants.

Lesser micro-organisms are present in simple composting. Hence, they solublize comparatively less amount of these minerals by their enzymatic activity. But in vermiculture larger population of micro-organisms are present inside the body of worms. These microbes not only solublize the minerals present in the organic waste but also absorb minerals from the surroundings to avail the plants. Animal dung act as pure organic matter but the mineral content varies. Lower nitrogen content is found in solid wastes from the city and peanut hulls.

Organic matter is available mostly in the form of animal dung and crop residues in the villages. The amount of crop residue generated from different crops varies. For eg., about 1800-2500 kg straw is generated from wheat field. The highest amount of crop residue (30 ton per hectare) is obtained from sugarcane field. All type of wastes can be used for vermicomposting but most of them require pre-composting before being fed to the worms. The moisture content of animal dung is high so it can be directly fed to the worms. But the micro-organisms present in it release gases like ammonia, methane and hydrogen sulphide due to which the worms do not enter inside such an atmosphere. Therefore, it is said that dung should be cooled before being fed to the worms. Crop residues do not release any gas but their outer covering is so hard that it can't be directly fed to the worms. Therefore, it is customary to moisten and soften the crop residues for some time. The horticultural waste and garbage also need a short duration of pre-decomposition before being fed to the worms.

Organic material management was not a problem until vermicomposting was done on small scale (5-10 kg per month). But it became compulsory with the advent of large scale vermicomposting. It is very important to evaluate the availability of waste material before establishing a vermicomposting unit. Often the objective of V unit establishment is the disposal of waste materials. Vermicomposting units are being established in big hotels, canteens and food processing units. It provides dual benefit, first the waste is disposed and secondly profit is also earned.

Sometimes the waste such as dry and hard wastes cannot be directly fed to the worms without prior decomposition. Several methods for the primary decomposition of these wastes have been developed. These methods include physical, chemical and biological processes. The wastes may require one or more methods for their decomposition of the wastes. Firstly, there should not be any toxic gases or harmful micro-organisms during primary decomposition. Secondly, the vermicasting produced should be of higher quality. Biological senitizers can be used to remove odour, gases, mosquitoes, flies and micro-organisms.

Now-a-days methods which increase the nutrient content of wastes having lower amount of nutrients. Crop residues have higher cellulose content which can be degraded by the biological processes carried out by micro-organisms so that the nutrient and mineral content of the wastes also increases. Several chemicals such as rock phosphate, lime stone, granite dust, gypsum, etc. have been found to be very effective in the primary degradation processes. Wastes from animal butchery, fish industry, edible oil factories can also be added to improve the mineral content of such wastes. In short, all types of wastes can be added and used for vermicomposting process with certain precautions.

CHAPTER IX

ADVANTAGES / DISADVANTAGES OF VERMICOMPOSTING

Earthworms and their vermicompost can induce excellent plant growth and enhance crop production. Edwards and Burrows (1988) reported that vermicompost consistently improved seed germination, enhanced seedling growth and development, and increased plant productivity significantly. Vermicompost increases the productivity of cereal crops such as wheat and rice as compared to the chemical fertilizers. Palaniswamy (1996) found that earthworms and its vermicompost improve the growth and yield of wheat by more than 40 %.

Reddy and Ohkura (2004) studied the agronomic impacts of vermicompost on sorghum (Sorghum bicolour). They compared it with normal compost and chemical fertilizers (N + P_2O_5). Sorghum grown with vermicompost showed significantly higher growth, viz., root length, number of leaves, plant height and shoot biomass over the normal compost and also over the chemical fertilizers. Guerrero (2010) reported the impacts of vermicompost on growth of Zea mays. Vermicompost also increases the production of horticultural crops Buckerfield and Webster (1998) reorted that vermicompost increases the yield of grape by two-fold as compared to chemical fertilizers. Vermicompost appears to be superior to conventionally produced compost. It can be used as

❖ An inoculant in the production of compost;

❖ A high quality animal feed;

❖ Sources of supplemental income.

ADVANTAGES OF VERMICOMPOST

THE SOCIAL SIGNIFICANCE

NUTRIENT CONTENT

Organic foods contain higher dry matter and mineral content as compared to conventional foods. Bourn and Prescott (2002) summarized a number of studies that compared the effect of inorganic and organic fertilizers on the nutritional value of crops. They concluded that the results were too variable to provide any definitive conclusions concerning the effect of fertilizer type on mineral and vitamin. Worthington (2001) compared the Organic and Conventionally Grown Foods such as fruits, vegetables, and grains. Researchers compared a combination of 41 previous studies, organic fertilizers and methods of production of crops vs. conventional farming methods and affects on foods. Nutrient content and nitrate levels in foods were also measured. Results showed that organic foods contained :

- ❖ 27% more vitamin C
- ❖ 21.1% more iron
- ❖ 29.3% more magnesium
- ❖ 13.6% more phosphorus
- ❖ Lower nitrate levels found in organic foods

Organically produces food products have higher nutrient content. The ratio of nitrate levels in conventional foods relative to organic foods ranged from 97 to 819%. In another study of conventional vegetables and organic vegetables, the later was found to contain three times less nitrate. The organic milk contains more nitrate than conventional milk's (Magkos et al, 2003). Woese et al. (1997) also reported in teir review that conventionally cultivated or minerally fertilised vegetables normally have far higher nitrate content than organically produced or fertilised vegetables.

Studies also show that that organic production methods lead to increases in nutrients, largely organic acids and polyphenolic compounds, many of which have potential human health benefits as antioxidants. Two hypotheses have been put forward to explain the possible increase in organic acids and polyphenolics in organic versus conventional foods. The first hypothesis considers the impacts of different fertilization practices on plant metabolism. According to this hypothesis synthetic fertilizers frequently make nitrogen more available for the plants than do the organic fertilizers and hence may accelerate plant growth and development. Therefore, plant resources are allocated for growth purposes, leading toa decrease in the production of plant secondary metabolites such as organic acids, polyphenolics, chlorophyll, and amino acids. The second hypothesis considers the responses of plants to traumatic conditions such insects, weeds, and plant pathogens. As organic production methods do not allow the use of insecticides, herbicides, and fungicides, it lays greater stresses on plants to devote greater resources toward the synthesis of their own chemical defense mechanisms. This can be attributed to their production in plant defense (Asami et al., 2003). Studies using organically and conventionally cultivated strawberries demonstrated that extracts from organic strawberries showed higher anti-proliferative activity against colon cancer and breast cancer cells than did extracts from conventional strawberries (Olsson et al., 2006). Suhane (2007) studied the chemical and biological properties of soil under organic farming using various types of composts and chemical farming using chemical fertilizers, i.e., urea (N), phosphates (P) and potash (K).

Table 12 : Farm soil properties under organic farming and chemical farming

Chemical and biological properties of soil	Organic farming (Use of composts)	Chemical farming (Use of chemical fertilizers)
Availability of nitrogen (kg/ha)	256.0	185.0
Availability of phosphorus (kg/ha)	50.5	28.5
Availability of potash (kg/ha)	489.5	426.5
Azatobacter (1000/gm of soil)	11.7	0.8
Phospho bacteria (100,000/kg of soil)	8.8	3.2
Carbonic biomass (mg/kg of soil)	273.0	217.0

Source: Suhane (2007)

Studies by Agarwal (1999) also found that the NPK value of vermicompost processed by earthworms from the same feedstock (cattle dung) significantly increases by 3 to 4 times. It also enhances several micronutrients.

Table 13 : NPK value of vermicompost compared with conventional cattle dung compost made from cattle dung

Nutrients	Cattle Dung Compost	Vermicompost
N	0.4-1.0%	2.5-3.0%
P	0.4-0.8%	1.8-2.9%
K	0.8-1.2%	1.4-2.0%

Source: Agarwal (1999); Ph. D Thesis, University of Rajasthan, India

Winter and Davis (2006) reported the levels of various nutrients present in different food products studied.

Table 14 : Nutrients levels in food products grown organically

Food Products	Nutrients Studied	Results
Cabbage, Spinach and Onion	Flavonoids	Higher levels of flavonoids
Peach and Pear	Polyphenoloxidase Enzyme Activity, Total Phenolics and Organic Acids	Both had higher levels of phenolic and polyphenoloxidase; organic peach had higher citric and ascorbic acids
Corn and Strawberry	Phenolics and Ascorbic Acids	Higher levels of phenolics and ascorbic acids
Tomatoes	Vitamin C, Carotenoids and Polyphenols	Higher levels of vitamin C Carotenoids and polyphenols
Grapes	Polyphenoloxidase and Diphenolase Enzymes	Polyphenoloxidase did not differ; diphenolase activity was 2 times higher
Apples	Phenolics	Higher phenolics in pulps

Source: Winter and Davis (2006)

Several researches show that the nitrogen content increase with time on using vermicompost in the soil structure.

- ➢ Fine sand particles cling to the mucous on the surface of vermicompost. This results into the aggregation of soil particles.
- ➢ It increases the water holding capacity of the soil.
- ➢ It provides a regular supply of moisture and thus promotes the decomposition of organic matter.
- ➢ Micro-organisms present in the vermicompost feed on this organic matter further decomposing it.
- ➢ These micro-organisms absorb minerals from atmosphere and avail it to the plants.

CHEMICAL-FREE FOOD

Organically grown fruits and vegetables have been found to be highly nutritious and have more beneficial nutrients, such as antioxidants, than their chemically grown counterparts (Anonymous, 2000). In a ten-year comparative study Mitchell (2007) reported levels of flavonoids quercetin and kaempferol in organic tomatoes (115.5 and 63.3 mg per gram of dry matter) were 79 % and 97 % higher than those in chemically grown tomatoes (64.6 and 32.06 mg per gram of dry matter) respectively. The levels of flavonoids also increased over time in samples of tomatoes treated organically. Studies indicate that organic foods are high in organic acids and poly-phenolic compounds many of which have potential health benefits like antioxidants (Winter and Davis, 2006). In addition, organic products have less nitrates than their chemical counterparts. Heaton (2001) found 14 studies showing average 50 % lower nitrates in organically grown crops. Shankar and Sumathi (2008) reported significantly higher 'nitrates' in chemically grown tomatoes. The reduction in free amino acids by organic fertilizers is beneficial for crops. Aphids feeding on plants use this as a source of protein.

REDUCED RISK OF SOME CANCERS

Studies show that organic foods can reduce the risks of cancer in humans. Extracts from organic strawberries showed higher anti-proliferative activity against colon cancer and breast cancer cells as compared to the extracts from conventional strawberries (Olsson et al., 2006). Tomato provides a balanced mixture of minerals and antioxidants, including vitamin C, total carotene and lycopene. Lycopene has been found to have preventive effects on prostate cancer in human beings. Significantly higher lycopene was reported in organically grown tomato by Lumpkin (2005).

SECONDARY METABOLITES IN PLANTS

Secondary metabolites like glucosinolates, glycoalkaloid, flavonoids, carotenoids and sulphur compounds can quench free radicals ans therefrore act as anti-proliferative agents. They promote detoxifying enzymes and induce differentiation of cancer cells. They inhibit metastasis, tumour blood vessel formation and stimulate the human immune system (Heaton, 2001). Chemical pesticides reduce the needs of plants to produce these beneficial secondary metabolites for their protection.

EFFECT ON SOIL STRUCTURE

SOIL FERTILITY

The vermicompost with a relatively high content of humus-like compounds, active micro organisms and enzymes, greatly contribute to the enhancement of the biochemical fertility of soils degraded by intensive – cultivation, pollution or natural causes (Perucci, 1992). The casts of earthworm is one of the most useful and active agent in introducing suitable chemical, physical and microbiological changes in the soil and, thereby, directly increasing the fertility and crop producing power in the soil (Joshi and Kelkar, 1951).

IMPROVEMENT OF CATION EXCHANGE CAPACITY

It also increases the cation exchange capacity of soil. It leads to higher adsorption of positively charged cations such as calcium (Ca), magnesium (Mg), potassium (K) and sodium (Na) by soil. This ultimately increases the availability of these minerals in the soil for plant uptake.

PREVENTION OF EROSION

Vermicompost and earthworms reduce the bulk density of the soil, prevent soil compaction and improve root growth, drainage and infiltration. This also reduces surface crusting and allows better infiltration of rainfall water. Decrease in rate of infilteration leads to increased soil erosion. It keeps the soil moist for longer duration.

REMOVES SOIL SODICITY AND SALINITY

A soil is regarded as sodic when exchangeable sodium (Na) is higher than 6 % and the pH is greater than 8.5. Gypsum is applied to reduce the sodicity and salinity by increasing the amount of exchangeable calcium in the soil. Compost can help in spread of gypsum at faster rates in the soil by increasing the soil porosity. Worms ingest soil and gypsum and the vermicastings thus produced results in fast and thorough spread of gypsum deep into the soil profile.

MAINTAINS OPTIMAL PH VALUE OF SOIL

Vermicompost also decreases the acidity and increases the soil pH. Thus it helps in maintaining an optimal pH of soil.

INCREASES WATER HOLDING CAPACITY OF SOIL

Earthworms act as natural plougher of the land and increase the number of pores in the soil. This leads to increase in water holding capacity of soil. Vermicompost retains moisture and maintains evaporation losses to a minimum and acts as an absorbent of atmospheric moisture due to the presence of colloidal materials, i.e., the earthworm mucus. It works as a kind of micro-dam storing hygroscopic and gravitational water. This ultimately leads to increase in productivity of the soil.

HUMIC SUBSTANCES

Vermicomposts consist of humified earthworm feaces and organic matter. Albanell et al., 1988; Petrussi et al., 1988; Senesi et al., 1992 reported that vermicomposts originating from animal manure, sewage sludges or paper-mill sludges contain large amounts of humic substances. Chen and Aviad (1990) support positive growth effects of the effects of humic substances on plant growth, under adequate mineral nutrition conditions. Humic acid application increases dry weights of shoots, roots, and nodules of soybean, peanut, and clover plants (Tan and Tantiwiramanond, 1983); vegetative growth of chicory plants (Valdrighi et al., 1996); and induced shoot and root formation in tropical crops grown in tropical crops grown in tissue culture

(Goenadi and Sudharama, 1995). Humic acids significantly stimulate plant growth as compared to that produced by mineral nutrients. The humic acids in humus are essential to plants in four basic ways –

1. Enables plant to extract nutrients from soil;

2. Helps to dissolve unresolved minerals to make organic matter ready for plants to use;

3. Stimulates root growth; and,

4. Helps plants overcome stress (Kangmin et al., 2010).

THE ECONOMIC SIGNIFICANCE

CONVERSION OF WASTE TO THE BEST

Vermicomposting converts useless waste to useful biofertilizers of high economic value. It reduces the cost of waste handling, transportation and their treatment. Landfills are used in most parts of the world for handling wastes. It costs a lot and also pollutes the land used for this purpose and the nearby water bodies as well as ground water. Vermicomposting prevents us from this economic loss.

LOW COST OF PRODUCTION

The cost of food production can be significantly reduced to about 50-60 % as compared to the conventional farming with chemical fertilizers. The raw material used is available absolutely free of charge and the food produced will be a chemical-free organic food. Vermicompost also reduces the cost of pesticides and herbicides. The worms build up the soil's physical, chemical and biological properties as well. vermicompost possesses higher water holding capacity and can therefore reduce the cost of irrigation too.

THE BIOLOGICAL SIGNIFICANCE

REMOVING DEAD ORGANIC MATERIAL

Earthworms are one of the most efficient biological agents for removing the dead organic materials and composting them to usable organic fertilizer, vermicompost.

Naturally, they reside in soil, near the roots of plants. Hence, they can directly provide the required nutrients to the plants. Composting can also be done by human near their fields by using the wastes from their farms, fields, kitchen, animal excreta etc.

ACCESSIBILITY OF NUTRIENTS TO THE PLANT

Earthworms increase the accessibility of nutrients to the plants. A study showed that earthworms take organic matter from the surface of the soil and migrate it to the upper soil horizons. This ultimately takes nutrients such as N and P and put it proximal to plants roots, increasing their availability.

SEED DISPERSAL

Earthworms also increase seed dispersal. A study showed a positive correlation between the number of earthworms present and the number of seedlings. This study showed that earthworms inadvertently eat plant seeds. The worms continue to burrow through the soil as these seeds pass through their digestive system. The seeds are untimely excreted within the soil. Not only is this a great seed dispersal system for plants, the seeds are also excreted in a worm cast which is a nutrient rich substance for the seed. Worms have been found to be sensitive to pesticides and herbicides, and thus have been used as bio-indicators of healthy or poor soils. Another study showed that *E. fetida* can be used as a bio-indicator of high concentrations of heavy metals such as copper and lead in the soil.

PESTICIDE RESIDUE

Pesticide residues, naturally occurring toxins, nitrites, and polyphenolic compounds exert an increased risk of certain diseases, such as some cancers, coronary heart disease. Organic foods can limit our exposure to pesticides and their ill-effects. Baker and others (2002), Pussemier and others (2006) reported that pesticide residue presence in conventional foods compared to organic foods

- ❖ PDP: 2.3 times greater
- ❖ CDPC: 4.8 times greater
- ❖ CU: 2.9 times greater
- ❖ Belgian: 4.1 times greater
- ❖ Pesticide residue was lower in organic produce than conventional

Table 15 : Comparison of pesticide residue presence in conventional and organic foods.

	USDA Pesticide Data Program	CDPR Market Place Surveillance Program	Consumers Union	Belgium
Conventional	73%	31%	79%	49%
Organic	23%	6.5%	27%	12%
Ratio	3.2	4.8	2.9	4.1

Sources : Baker and others (2002), Pussemier and others (2006)

MICROBIAL SAFETY

Synthetic fertilizers are not used in organic farming. İnstead of synhetic fertilizer, animal manure, vegatable waste (gren manure), marine waste products are used. Therefore, Organic foods contain more pathogenic microorganisms such as Salmonella, Listeria and Escherichia coli as compared to conventional foods (Winter and Davis, 2006). Micro-organisms are responsible for transforming, releasing and cycling of nutrients and essential elements. These micro-organisms are also essential for converting nutrients into their plant available forms and also for facilitating nutrients uptake. Soil microbes also create glue like substance that sticks soil particles together, creating pore spaces that make good soil structure decreasing soil hardness. But in vermicomposting, the worms eat the organic matter and consume the bacteria, as well as harmful nematodes, weed seeds, and pathogenic fungi. The worm's digestive system breaks down the bacteria and other harmful substances and excretes the undigested matter. This is further consumed by other microorganisms present in the bin. This reduces the risk of microbial infection. Level of pathogens namely E.coli, Faecal Coliforms and Salmonella spp. reduces upto 99.9%. Vermicomposting exhibits greater pathogen reduction than that of conventional composting. All three of these pathogens are not obligate aerobes. The obligate aerobes namely *Pseudomonas spp.*, *Zoogloea spp.*, *Micrococcus spp.* and *Aebromobacter spp.* have evolved to process nutrients and reproduce at the highest efficiency in aerobic conditions and so the pathogens are excluded from nutrients and space as the obligate aerobes continue to increase under ideal conditions.

WEED CONTROL

The worms generally digest the added weed seeds as well. Therefore, there is no chance of further spread of weeds in the fields by the use of vermicompost. Researchers around the globe are more interested in harmful microbes as compared to the advantageous species. Therefore, the following data is available regarding them based on the process of manufacturing of vermicompost.

➢ diarrhea causing bacteria die within one hour at 55^0C and 20 minutes at 60^0C.

➢ typhoid causing bacteria, *Salmonella typhi* cannot survive above 46^0C.

➢ Amoebic dysentery causing bacteria dies within few minutes at $45\ ^0C$ and few seconds at $50\ ^0C$.

➢ Tapeworm and hukeworm die at $55\ ^0C$.

➢ Polio and tuberculosis causing agents also do not survive under vermicomposting conditions.

During initial degradation of waste of 1 m size the temperature rises to more than $70\ ^0C$ in about 10 days. Hence, it can be presumed that almost all the harmful micro-organisms are destroyed during the process of vermicomposting. In addition to this, the mucous produced by the worms also helps in destroying them. Mucous is a sticky substance which stops the supply of food to the plants.

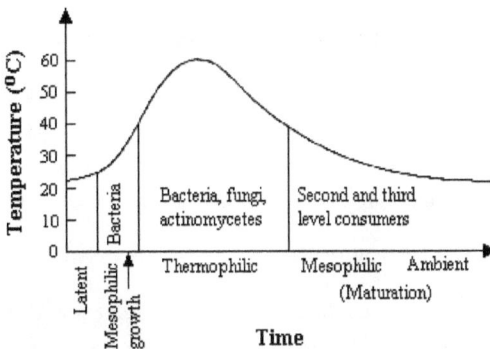

Patterns of temperature and microbial growth in compost piles (Polprasert, 1996).

Table 16 : Different types microorganisms present in compost piles.

Number of Bacteria	More Than 10^{10}
Number of actinomycetes, azotobacter, rhizobium, phosphate solublizer and nitrobacter	About 10^6 - 10^7
Giberellins (G.A.), auxins (I.A.A.) and cytokinins (I.P.A.)	Adequate amount
Fungi	Many types of useful fungi

ABILITY TO SUPPRESS DISEASE

Vermicompost possesses the ability to protect plants against various diseases. It is assumed that the high levels of beneficial microorganisms such as *Pseudomonas spp.*, *Zoogloea spp.*, *Micrococcus spp.* and *Aebromobacter spp.* in vermicompost protect plants by out-competing pathogens for available resources. Vermicompost occupy all the accessible sites around plant roots and prevent the access of pests. This assumption is based on the concept of the SOIL FOODWEB, a soil-ecology-based approach given by Dr. Elaine Ingham of Corvallis, Oregon. Edwards and Arancon, 2004 reported that worm castings sometimes repel hard-bodied pests. The production chitinase enzyme by the worms might be responsible for this as it breaks down the chitin in the insects' exoskeleton.

HIGHER BIO-AVAILABILITY OF NUTRIENTS FOR PLANTS

Vermicompost contains nutrients in forms such as nitrates (N), phosphates (P), soluble potassium (K), and magnesium (Mg) and exchangeable phosphorus (P) and calcium (Ca) which are easily available for plant uptake. Vermicomposts is porous and hence has larger surface area for microbial activities (Arancon and Edwards, 2006).

RICH IN GROWTH HORMONES

Vermicompost stimulates plant growth due their higher content of plant growth hormones. It improves seed germination, enhances seedling growth and development and increases plant productivity much more than would be possible from the mere conversion of mineral nutrients into plant available forms. Neilson (1965) and Tomati et al., (1987) reported that vermicompost contains growth promoting hormone such as auxins, cytokinins and flowering hormone gibberellins secreted by earthworms.

TOXIC CHEMICALS FREE BIOFERTILIZER

Sinha and Valani (2011) reported that earthworms effectively bio-accumulate or biodegrade several organic and inorganic chemicals including heavy metals, organochlorine pesticide and polycyclic aromatic hydrocarbons (PAHs) residues in the medium in which it inhabits.

BIO-PESTICIDAL ACTION

Arancon et al. (2002) and Chaoui et al. (2002) have reported the ability of earthworms and vermicompost to protect plants against various pests and diseases either by suppressing or repelling them or by inducing biological resistance in plants to fight them or by killing them through pesticidal action. Plants grown with vermicompost are less susceptible to a number of arthropod pests and sustain significantly lower pest populations due to balanced nutrients and greater microbial and faunal diversity compared to chemical fertilizers.

The mechanisms of vermicompost-mediated plant defenses against insect pests has not been interpreted properly but the probable reasons may be due to the presence of

> ➤ antagonistic microbes
> ➤ fungivorous and bacterivorous nematodes
> ➤ insect-parasitic organism such as the entomopathogenic nematodes (EPN) in vermicompost.

Vermicompost works to protect crops in the following ways

Induction of Biological Resistance in Plants

Vermicompost contains some antibiotics and actinomycetes increases the biological resistance of the crop plants against pest and diseases. Sinha and Valani (2011) found that the pesticide spray was significantly reduced where earthworms and vermicompost were used in agriculture. This resistance might be offered by the interactions between the microbial populations in vermicompost with plant roots.

Repelling Crop Pests

Higher concentration of nitrogen attracts aphids more due to more free amino acid availability. Crops fertilized organically are less affected by aphids. Organically grown plants have higher mechanical strengths of cell walls and decreased water contents in plant tissues. This favours protection against aphids.

THE ENVIRONMENTAL EFFECTS

Aristotle called worms the intestines of the earth. Charles Darwin wrote a book on worms and stated that worms surpass any other creatures in their contribution to life on earth. Vermicomposting and vermiculture are eco-friendly processes and have great prospective in sustainable agriculture. It is practiced in countries such as Cuba, India, Argentina, and Australia. It is beneficial as a fertilizer as well as in the form of a business.

Two essential aspects of composting include recovering organic residuals and the benefits of adding compost to the soil to restore the organic content and rebuild the soil food web. This is really beneficial from the environmental point of view. It involves the composting of wastes, returning nutrients to the soil and thus the reducing the use of artificial fertilizers.

However, there are also negative impacts of composting on the environment. Gases released such as carbon dioxide, volatile organic compounds from improperly maintained compost piles and bacteria and fungi are negative effects associated with composting. The release of methane and carbon dioxide also contribute to the problem of greenhouse gases in the atmosphere. Poorly operated composting also causes unpleasant odors. Mixed MSW and sewage sludge contain higher levels of heavy metals than do kitchen or yard wastes. Therefore, they pose greater risks. This is a serious apprehension, and controlling should be done through :

- analysis of composts;
- development and enforcement of land application standards; and
- research and development.

Leachate production is also a serious problem. Leachate from water runoff and condensation at compost areas contain high biological oxygen demand (BOD) and phenols, a byproduct of the decomposition of the lignin in leaves. It may pose serious problems if leaches into the ground. Higher BOD can reduce the amount of dissolved oxygen in lakes and streams that is essential for aquatic life. Therefore, one should avoid the discharge to water and all leachate should be directly absorbed in sand or soil.

REPLACING THE ENVIRONMENTALLY DESTRUCTIVE AGROCHEMICALS

Huge toxic wastes, greenhouse gases and pollution are generated at all stages of production of chemical fertilizers. Vermicomposting involves no such burden on the nature. In fact, it is a totally natural phenomenon.

REDUCING IN THE EMISSION OF GREENHOUSE GASES

Vermicompost is produced from waste materials of society. Composting of wastes by worms is proving to be economic and environment friendly technology. It is almost odourless process giving the detoxified and disinfected end product. Landfills are being reduced in number. They emit large amount of greenhouse gases.

CARBON SEQUESTERING IN SOIL

A large amount of carbon is held in the soil. This is known as soil organic carbon (SOC). It amounts to about 41 % whereas the soil inorganic carbon amounts to 23 %. This is about three times of the atmospheric carbon pool as CO_2 which is 20 %. Since 7000-10,000 yrs ago the balance between the SOC and atmospheric carbon has been changing. Aggressive ploughing and tillage leads to the loss of SOC. It has increased the atmospheric carbon. This increases greenhouse effect and leads to global warming and climate change. Soil erosion is also a major cause of the loss of SOC. Use of fossil fuels has also intensified the situation.

Robbins (2004) reports that about a third (33.3 %) of the increase of atmospheric carbon over the last 150 years is thought to have come from agriculture. Carbon sequestration is a process of putting more carbon back into the soil through sustainable agricultural practices mainly organic farming by the use of composts. Compost use in farms would 'sequester' huge amounts of atmospheric carbon (CO_2) and bury them back into the soil, mitigate greenhouse gases and global warming. Vermicomposts are the composts prepared by the earthworms in the form of vermicasts. They and secrete more stable carbons in the form of humates to be retained in soil.

WATER QUALITY ISSUES

Previously it was believed that vermicomposting is a mesophilic process, so it does not destroy potentially dangerous pathogens. In recent years, strong evidence has been put forward that worms destroy pathogens, although the mechanism is still unknown. Dr. Elaine Ingham of Orange County Environmental Protection Division, Florida has found in her research that worms living in pathogen-rich material, when dissected, show no evidence of pathogens beyond the first five millimeter of their gut. In other words, something inside the worm destroys the pathogens, leaving the castings pathogen-free (Appelhof, 2003).

Vermicompost used in fields do not result in pathogen contamination of ground or surface waters. This adds earthworms to the soil. These worms again decompose the wastes and animal dung in the fields. This adds humus to the soil thus preventing water contamination by pathogens. Vermicompost, binds the soil particles, improves its structure, aeration and water holding capacity. Thus, it binds nutrients in the soil. This leads to decrease in nutrient run-off. Compost worms act as a part natural filtration systems.

CLIMATE CHANGE FACTORS

Climate change is one of the burning issues in environmental science. Farms and fields play role in climate change, mainly by the release of carbon from soils and the generation of methane gas from livestock and their manure. Vermicomposting can prove to be a solution to these problems. Vermicomposting leads to carbon sequestration. This is the process of locking carbon in the form of organic matter and organisms in the soil. More carbon is locked up in the soil than raw manure or inorganic fertilizer due to the stability of composts. Inorganic fertilizers lead to reduction in soil carbon content. Application of vermicompost increases the carbon content of the soil. The composting process is believed to be neutral with respect to greenhouse gas generation. The United States Environmental Protection Agency (US EPA) assessed the GHG impact of composting yard wastes a few years ago as part of a larger assessment of recycling and climate change. Their findings were that the composting process results in the same level of GHG emissions as if the materials were allowed to decay naturally, as on the forest floor. This can be attributed to the cut-off in various operations during the vermicomposting process. For eg: there is no need to transport the waste organic matter and the finished products to longer distances. It reduces the use of chemical fertilizers as well as the eutrophication process. It also reduces the emission of methane and nitrous oxide. Vermicompost contains higher levels of nitrogen. It is more efficient at retaining nitrogen. This can be attributed to the greater numbers of microorganisms present in the process.

BELOW-GROUND BIODIVERSITY

Biodiversity loss is an important issue of concern to the environment. Earthworms play an important role in counteracting the loss of biodiversity. Worms increase the numbers and types of microbes in the soil by creating favourable environment. Addition of vermicompost to the soil increases the microbial population.

NEGATIVE EFFECTS

INVASIVE SPECIE

Certain earthworm species act as invasive species. For eg, earthworms are an invasive species in the native northern hardwood forests of United States. They earthworms are introduced by human activity such as anglers dumping unused worms in waterways or on soil, construction and forestry road construction, and by vermiculturalists. Although earthworms are great for plants, a study showed that invading earthworms were correlated with a loss of understory plant species, and increase is soil carbon loss, and their effect of nutrient cycling.

IMPACT NUTRIENT CYCLING

The disturbance in soil structure can greatly affect nutrient cycling. Certain important nutrients such as nitrogen (N) and phosphorus (P) become less available, or end up lost via leaching (Hale et al. 2005; Frelich et al. 2006). It has been found to reduce ground vegetation.

EFFECT ON SOIL STRUCTURE

Loss of The O Horizon – It is the uppermost layer in the soil profile and consists primarily of fallen leaves and other organic matter.

Mixing of Organic Matter into the A Horizon – It follows the O horizon of soil.

When no earthworms are present, this organic matter breakdown generally takes place over the course of years – whereas the process is often greatly accelerated when earthworms are present.

LACK OF STUDIES

Threats imposed by more newer species in vermicomposting, such as *Amynthas sp* (Jumpers) has not yet been adequately assessed. More work needs to be done in this area. Different worm species have different requirements and thus have different impacts on the ecosystems. Different combinations of species can result in different impacts as well (Hale et al. 2005).

CHEMICAL TOXICITY

Large amount of heavy metals may be present in various wastes used in vermicomposting process. The waste materials may vary in terms of their content of heavy metals and their distribution of these metals in their various chemical forms. This variation depends on the waste production process, origin of the wastes, and the type and level of treatment undergone. The mobility of heavy metals in the soil is strongly determined by their chemical form in the waste materials.

The advantages and disadvantages of vermicomposting can be summarized as

ADVANTAGES

Soil texture

- ❖ It improves
  - soil texture
  - water holding capacity of the soil.
  - Soil aeration and
- ❖ It prevents soil erosion.
- ❖ It attracts deep-burrowing earthworms already present in the soil.
- ❖ Vermicompost contains earthworm cocoons and increases the population and activity of earthworms in the soil.

Plant growth

- ❖ It enhances sugar content and thus the quality of grains/ fruits.
- ❖ It adds plant hormones such as auxins and gibberellic acid to the soil.
- ❖ It enhances germination, plant growth, and crop yield.

- ❖ It improves root growth and structure.
- ❖ It adds enzymes such as phosphatase and cellulase to the soil.
- ❖ It provides macro- and micro- nutrients to the plants.
- ❖ It prevents nutrient losses and increases the use efficiency of chemical fertilizers.
- ❖ Vermicompost is free from pathogens, toxic elements, weed seeds etc.
- ❖ It contains valuable vitamins, enzymes and hormones like auxins, gibberellins etc.
- ❖ It restores microbial population,viz., nitrogen fixers, phosphate solubilizers etc.
- ❖ Microbial activity in worm castings is 10 to 20 times higher than in the soil.

Economic

- ❖ It creates low-skill jobs at local level
- ❖ Good quality manures can be prepared locally.
- ❖ It saves transportation
- ❖ It requires low capital investment and
- ❖ It is relatively simple technologies make vermicomposting practical for less-developed agricultural regions.

- ❖ Vermicompost requires no curing (as traditional composted materials do) as it is already populated with beneficial microorganisms.
- ❖ Vermicomposting is much faster than regular composting. It takes about 1 month whereas normally it might take 6 months
- ❖ Vermicompost is free flowing, easy to apply, handle and store and does not have bad odour.

Environmental

- ❖ Eco-friendly technique
- ❖ Organic manure is prepared from solid waste

- ❖ It decreases the use of fossil fuels
- ❖ It supports sustainable development principles
- ❖ Biowastes conversion reduces waste flow to landfills
- ❖ Waste recycling can be done on-site
- ❖ Its Production reduces greenhouse gas emissions such as methane and nitric oxide produced in landfills or incinerators
- ❖ It enhances the decomposition of organic matter in soil
- ❖ The vermicomposting process also reduces waste volume (up to 60%).
- ❖ It decreases the use of pesticides for controlling plant pathogens

DISADVANTAGES

- ❖ It is more complicated process than traditional composting
- ❖ It requires more labour
- ❖ It requires more space because worms are surface feeders but can't operate more than one meter thick material
- ❖ It is more susceptible to environmental factors, such as temperature
- ❖ It requires more start-up resources
- ❖ It requires higher cost

CHAPTER X

TROUBLESHOOTING

As mentioned earlier, worms like to live in a healthy environment and if they try to escape out of the bin, it indicates that problems exist.

Problems	Causes	Solutions
Foul odour	Too much air	smaller holes should be made
	Not enough air	bigger holes should be made
	Too much organic matter	lesser organic matter per feeding must be added
	Too acidic	Add some calcium carbonate and cut down on the amount of citrus peel and other acidic food waste
	Too wet	more drainage holes should be made
	Too dry	the compost should be moistened to add water
Dying worms	No food	food must be added
	No bedding for worms	compost must be harvest and bedding must be added
	Extreme temperatures	Moderate temperature must be maintained
Crawling away worms	Bin too dry	Thoroughly dampen bedding
Fruit Flies	Air holes too big	smaller holes should be made or organic matter should be burried under the bedding
Excess mold	Conditions too acidic	Cut back on acidic foods
Excess drainage	Poor ventilation Fluff bedding	add dry bedding
Too much water in food		Cut back on coffee grounds and watery scraps
Extreme temperatures		Move bin to 70.80^{0}F location

TIPS & WARNINGS

- **General rule of thumb** - 1 Kg of worms eats approx. 1/2 Kg of food waste per day.

- A records of types of bedding and food added to the bin should be maintained to track any problems.

- Banana peels and other fruit peels should be washed well before adding to the bin to clean off pesticides.

- In case of a population crash worms should be moved to another bin. Fresh newspaper or clean peat should be added as bedding. The original bin should be checked for excessive dampness, excessive dryness, lack of food, the wrong temperature or too much light.

- Chemically treated paper, such as shredded credit card receipts or thermal fax paper should not be added to the worm bin.

- Unchlorinated water should be used while spritzing the worm bin bedding.

- Placing bones, meat, dairy, oils, salty foods, grass and inorganic products should not be added to the worm bin.

- Materials like plastic bags, bottle caps, rubber bands, sponges, aluminum foil, glass, etc. should not be added in the bin.

- Pet wastes or biodegradable diapers should not be added to the bin because of the presence of harmful pathogens in faeces.

- Certain inks and dyes may alter the pH levels within the system.

- Worm bins should not be exposed to extreme temperatures.

- Insecticides should not be used around the worm bin.

- Garden soil should not be used as bedding for the worms.

- Fresh animal dung should not be added to the bin.

- Sanitation should be taken care of and hands should be washed properly after working with a worm bin.

CHAPTER XI

CASE STUDIES

VERMITECH 200 SYSTEM

It is used by the Medical University of South Carolina. It measures 18' x 7'. It processes approximately 200 pounds of cafeteria food waste each day. The system has a series of curved bars spanning the width of the bin Instead of a lid. When temperatures are cool, these support a thick cover. The composting chamber consists of two eight-foot sections. An air conditioner and hydraulic equipment occupy the last two-foot section. Feedstock is delivered to the bin through a vermitech shredder. Feedstock is combined with shredded cardboard. Then it is spread onto the surface of the bedding. The bottom layer of castings is removed after every 2 or 4 weeks by the hydraulic system. These are then used by the grounds department. The system has a payback time of three years. The floor is coated with acrylic so that castings can be cleaned up easily. The system also incorporates electricity, water and fans to control airflow. Start-up cost was $54,000 (including the building and supplies).

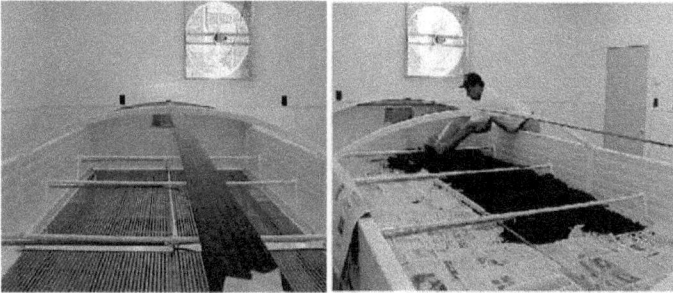

ORGANIC AGRICULTURE CENTRE OF CANADA, NOVA SCOTIA

Vermiculture Trial – Scott Farm

Jennifer Scott operates a small organic poultry operation at Centre Burlington, Nova Scotia. Organic Agriculture Centre of Canada (OACC) worked with Jennifer in an 18-month project. The aim of the project was to assess the opportunity for raising compost worms as a feed for her chickens. Organic grain is difficult to get and expensive in Nova Scotia. The worms provide high-quality protein necessary for the chickens and eliminate the need for importing grain.

Holdanca Farms Ltd.

This farm is operated by John Duynisvelt and is located near Wallace, Nova Scotia. The farm is run using organic methods, without any pesticides, commercial fertilizers or other restricted inputs. The farm produces free-range beef, poultry and pork. A large amount of manure is produced at the farm. Vermicomposting of this manure is done and the beef is sold locally.

GRIFFITH UNIVERSITY, AUSTRALIA

Potted Wheat Crops

This was designed to compare the agronomic impacts of vermicompost with conventional compost & chemical fertilizers on wheat crops. Wheat crops maintained very good growth on vermicompost & earthworms from the very beginning & achieved maturity in 14 weeks. Plants were greener and healthier over others, with large numbers of tillers & long seed ears were formed at maturity. Seeds were healthy and nearly 35-40% more as compared to plants on chemical fertilizers. More significant was that the pot soil with vermicompost was very soft & porous and retained more moisture (Sinha et al., 2010).

Potted Tomato Plants

This was designed to compare the agronomic impacts of vermicompost & worms with composted cow manure from market & chemical fertilizers on tomato plants. Tomato plants on vermicompost & vermicompost with worms maintained very good growth from the very beginning. Number of flowers and fruits per plant were also significantly high as compared to those on agrochemicals and conventional compost. Presence of earthworms in soil made a significant difference in flower and fruit formation in tomato plants. This was obviously due to more growth & flowering hormones (auxins and gibberlins) available in the soil secreted by live earthworms (Sinha et al., 2010).

ASHDEN AWARDS 2007 CASE STUDY

Country : India

State : Karnataka

The rural population of Karnataka mainly uses fuelwood for cooking. Due to this the pressure on land is increasing day-by-day. SKG Sangha is a not-for-profit organisation, founded in 1993 to promote development and the use of renewable energy technologies in South India. SKG Sangha encouraged the rural mass in supplying plants to large numbers of households to replace the use of fuelwood. 'Deenbandu' plants were built by local masons and labourers trained by SKGS. Local and central governments provide subsidies to the people. 43,000 plants were installed by 2007.

> ➤ Plants produce biogas by digesting cow dung, producing biogas which replaces fuelwood used and also kerosene for cooking.
> ➤ They save about 170,000 tonnes/year of CO_2.
> ➤ It also solves several health issues such as respiratory complaints, eye problems and headaches. It acts as a source of income generation for the rural masses.
> ➤ The vermicompost from biogas residue is sold for income.
> ➤ Thus, it has generated employment oppurtunities with lower capital requirement alongwith maintaining eco-friendly conditions.

SKG Sangha received the 2007 Ashden Award for its efforts.

GRAM LAXMI VERMICOMPOSTING INITIATIVE

Country : India
State : Karnataka
District : Sabarkantha

The Gram Laxmi initiative was started as a pilot project in 2011in 25 villages by the District Rural Development Agency (DRDA) to convert agricultural and animal waste into organic manure. In Hindi, Gram stands for village and Laxmi is the Goddess of money and prosperity. This initiative has been named as Gram Laxmi initiative because women are involved in the various operations of the vermicomposting units. It runs under Mission Mangalam- a Government of Gujarat (GOG) livelihood and poverty alleviation programme. Under the project, Gram Laxmi vermicomposting units are set up in villages and Self Help Groups of women are trained to run and manage these units. The agricultural and cattle waste from farms are collected and treated in the unit. Vermicompost is obtained from the unit and used as manure during farming. This vermicompost is sold to farmers at nominal rates as well as used for personal consumption on the women's farms.

Different centrally sponsored rural development schemes like the Total Sanitation Campaign (TSC), MGNREGA, National Rural Livelihood Mission (NRLM), Swaranjayanti Gram Swarozgar Yojana (SGSY), Backward Region Grant Fund (BRGF), Nirmal Gram etc. a one-time public contribution formation the establishment of Gram Laxmi units. The Gram Laxmi initiative has been scaled to 96 villages of Sabarkantha district. It successfully demonstrates

- ➤ The potential that rural areas have for developing indigenous and sustainable livelihood options.
- ➤ The use of localised and easily available raw materials,
- ➤ The pooling together of funds from well established national and state level schemes and
- ➤ The leveraging of existing network of local human resources (SHG women).

VERMICOMPOSTING ENTERPRISES FOR RURAL WOMEN

Country : India

State : Karnataka

District : Tumkur

Development Alternatives (DA) launched a project in mid-1996 to help underprivileged rural women develop vermiculture micro-enterprises. The project was initiated with support from the Council for Advancement of People's Action and Rural Technology (CAPART). The objective of this 18-month project was

- ➢ to help women from rural and peri-urban areas to set up micro-enterprises based on vermiculture technology.
- ➢ the improvement of soil fertility and increased crop productivity through ecological methods of farming.

It was carried out in three gram panchayats of Huliyar Hobli in Chikkanayakanahalli taluk of Tumkur district. The peri-urban areas included the small town of Bukkapatna in Sira taluk and Huliyar town in Chikkanayakanahalli. The project area was around 155 km north west of Bangalore.

The entrepreneurs were selected based on criteria such as levels of income and backwardness, availability of space, access to water and the willingness to spend time on the vermicompost training. The field staff of DA, simultaneously carried out studies on the availability of different types of organic wastes in the area, crops and use of manure, land holding patterns and related aspects. The training programme covered technical aspects of breeding earthworms, managing collection of organic wastes, application of vermicompost for various crops, managing the production system, accounting and marketing. Today, 25 women-run enterprises are functional and making good profit, in these areas.

VERMICOMPOST PRODUCTION UNIT PATNA

Country : India

State : Bihar

District : Patna

Understanding the benefits of using vermicomposting Bihar government has started encouraging public for the establishment of vermicomposting unit with a capacity of 3000 MT per annum as a business as well as its use. Bihar government is offering 50% one-time capital subsidy (maximum Rs. 25 lakh) to entrepreneurs and Rs. 3/- per kg continuous subsidy to farmers for purchase of vermicompost.

VERMICULTURE IN INDIA

Agricultural Bioteks

Country : India

State : Maharatra

Agricultural Bioteks (India) was formed and established in 1985 as a small plant to manufacture vermicompost from agricultural waste. It is an organization which has initiated both commercial and educational ventures to promote vermiculture. Currently it produces 5,000 tons of vermicompost annually. About 2,000 farmers and horticulturists have adopted vermicomposting. Maharashtra Bioteks and the India Department of Science and Technology promoted the adoption of vermicompost technology in 13 states in India in 1991-92. About 1,000 farmers have reduces their use of chemical fertilizers by 90% by using vermicompost as a soil amendment for growing grapes, pomegranates and bananas. Similar work is underway on mangoes, cashews, coconuts, oranges, limes, strawberries and various vegetable crops.

WEED UTILIZATION FOR VERMICOMPOSTING

Country : India

State : Madhya Pradesh

District : Jabalpur

Weed biomass is one of the most abundant sources of organic matter and plant nutrients. They have not received much attention. The favourable climatic condition of the North Eastern region in general and Assam in particular leads to the production of huge weed biomass of diverse species such as Ipomea carnea, Mikania micrantha, Cassia occidentalis, Cassia tora, Lantana camara, Ageraum conyzoides, etc.

Ageraum conyzoides

Cassia tora

Lantana camara

Mikania micrantha

This technology of vermicompost production from weed biomass is being provided to the farmers/farm, women/educated youth of the region in general and Assam in particular through on-station and on-farm training and demonstrations. More than 2,500 farmers/farm women/educated youth have received training so far. The technology of vermicompost production from weed biomass has been well accepted by the farming community and already small units of vermicompost production have become popular in different parts of the State.

REFERENCES

1. Agarwal, S. (1999). Study of vermicomposting of domestic waste and the effects of vermicompost on growth of some vegetable crops; Ph. D Thesis (Supervisor: Dr. Rajiv K. Sinha) Awarded by University of Rajasthan, Jaipur, India.

2. Albanell, E.; Plaixats, J. CaBrero, T. (1988). Chemical changes during vermicomposting (Eisenia fetida) of sheep manure mixed with cotton industrial wastes. Biology and Fertility of Soils. 6: 266-269.

3. Anonymous. (2000). Organic food is far more nutritious: Newsletter of the National Assoc. Of Sustainable Agriculture Australia (NASAA); Feb. 10, 2000.

4. Anonymous. (2004). Market driven ecoenterprises for livelihood security. Market driven ecoenterprises for livelihood security. pp. 16.

5. Appelhof, Mary. (1997). Worms eat my garbage. Flower Press, Kalamazoo, Michigan.

6. Appelhof, M. (2003). "Notable Bits". In WormEzine, 2 (5).

7. Arancon, N.Q.; Edwards, C.A., Lee, S. (2002). Management of plant parasitic nematode population by use of vermicomposts; Proceedings of Brighton Crop Protection Conference on Pests and Diseases, 8B-2: 705-716.

8. Arancon, N.Q.; Edwards, C.A. (2004). Vermicompost can suppress plant pest and disease attacks, Biocycle, 51-53.

9. Arancon, N.Q.; Edwards, C.A. (2006). Effects of vermicompost on plant growth; In: uerrero, R.D. and Guerrero-del Castillo, M.R. (Eds): Vermitechnologies for Developing Countries, Philippine Fisheries Association, Laguna, Philippines., pp. 32-65.

10. Asami, D.K.; Hong, Y.J.; Barrett, D.M.; Mitchell, A.E. (2003). Comparison of the total phenolic and ascorbic acid content of freeze-dried and air-dried marionberry, strawberry, and corn grown using conventional, organic, and sustainable agricultural practices. J. Agric. Food Chem. 51:1237-1241.

11. Atiyeh, R.M.; Subler, S.; Edwards, C.A.; Bachman, G.; Metzger, J.D.; Shuster, W. (2000). Effects of vermicomposts and composts on plant growth in horticultural container media and soil. Pedo biologia. 44: 579–590.

12. Baker, B.P.; Benbrook, C.M.; Groth, E.; Benbrook, K.L. (2002). Pesticide residues in conventional, integrated pest management (IPM)-grown and organic foods: insights from three U.S. data sets. Food Addit. Contam. 19: 427-446.

13. Barik, T.; Panda, R.K.; Dash, S.; Nayak, M.P.; Sontakke, B. (2005). Effect of different farm wastes on vermicomposting. J. Applied Zoological Researches. 16(1): 106-107.

14. Baxter, G.J.; Graham, A.B.; Lawrence, J.R.;Wiles, D.; Paterson, J.R. (2001). Salicylic acid in soups prepared fromorganically and non-organically grown vegetables. Eur. J. Nutr. 40: 289-92.

15. Bintoro, G.; Blount, C.; Edwards, C.A. (2002). The growth and fecundity of Eisenia fetida (Savigny) in cattle solids precomposted for different periods. In Pedobiologia, 46: 15-23.

16. Biradar, A.P.; Nandihalli, B.S.; Jagginavar, S.B. (2000). Influence of seasons on the biomass of earth worms and vermicompost production. Kar. J. Agric. Sci.13 (3): 601-603.

17. Bogdanov, P. (1996). Commercial Vermiculture: How to Build a Thriving Business in Redworms. Vermico: Merlin, Ore.

18. Bouche, M. (1987). Emergence and development of vermiculture and vermicomposting from a hobby to an industry, from marketing to a biotechnology from irrational to credible practices. On Earthworms, Selected Symposia and

Monographs (eds. Bonvicini Pagliai, A.M., & Omodeo, P.), Mucchi Editore, Modena, Italy, pp. 519.

19. Bourn, D.; Prescott, J. (2002). A comparison of the nutritional value, sensory qualities and food safety of organically and conventionally produced foods. Critical Review of Food Science & Nutrition, 42: 1-34.

20. Brown, G.G. (1995). How do earthworms affect microfloral and faunal community diversity? Plant Soil, 170 (1), 209–231.

21. Buchsbaum, R.; Buchsbaum, M.; Pearse, J.; Pearse, V. (1987). Animals Without Backbones. Third edition. University of Chicago Press: Chicago, Ill.

22. Buckerfield, J.C.; Webster, K.A. (1998). Worm-worked waste boost grape yield: prospects for vermicompost use in vineyards, The Australian and New Zealand Wine Industry Journal., 13: 73-76.

23. Bulluck, L.R.; Brosius, M.; Evanylo, G.K.; Ristaino, J.B. (2002). Organic and synthetic fertility amendments influence soil microbial, physical and chemical properties on organic and conventional farms. Applied Soil Ecology, 19: 147-160.

24. Callaham, M.A.Jr.; Gonzalez, G.; Hale, C.M.; Lachnicht, S.L.; Zou, X. (2006). Policy and management responses to earthworm invasions in North America. Biological Invasions. 8: 1317-1329.

25. Chen, Y.; Aviad, T. (1990). Effects of humic substances on plant growth. Humic Substances in Soil and Crop Sciences: Selected Readings. ASA and SSSA, Madison, Wisconsin, USA, 161-186.

26. Carbonaro, M.; Mattera, M. (2001). Polyphenoloxidase activity and polyphenol levels in organically and conventionally grown peach (Prunus persicaI L., cv. Regina bianca) and pear (Pyrus communis L., cv.Williams). Food Chem. 72: 419-24.

27. Carbonaro, M.; Mattera, M.; Nicoli, S.; Bergamo, P.; Cappelloni, M. (2002). Modulation of antioxidant compounds in organic vs. conventional fruit (peach, Prunus persica L., and pear, Pyrus communis L.). J. Agric. Food Chem. 50: 5458-5462.

28. Caris-Veyrat, C.; Amiot, M.J.; Tyssandier, V.; Grasselly, D.; Buret, M.; Mikolajczak, M.; Guilland, J.C.; Bouteloup-Demange, C.; Borel, P. (2004). Influence of organic versus conventional agricultural practice on the antioxidant microconstituent content of tomatoes and derived purees; consequences on antioxidant plasma status in humans. J. Agric. Food Chem. 52: 6503-6509.

29. Chaoui, H.I.; Edwards, C.A.; Brickner, A.; Lee, S.S.; Arancon, N.Q. (2002). Suppression of the plant parasitic diseases: Pythium (damping off), Rhizoctonia (root rot) and Verticillium (wilt) by vermicompost. Proceedings of Brighton Crop Protection Conference on Pest and Diseases.

30. Chinnappa Reddy, B. V.; Subba Reddy, P.N.; Kale, R. D. (2007). Economic impact and production efficiency of vermicompost use in agriculture: methodological approaches, Agric. College, University of Agric. Sci., VC Farm, Mandya.

31. Ciavatta, C.; Govi, M.; Sequi, P. (1993). Characterization of organic matter in compost produced with municipal solid wastes: an Italian approach. Compost Sci. Util., 1 (1), 75-81.

32. Civilini, M.C.; de Bertoldi, M.; Sebastianutto, N. (1996). Composting and selected microorganisms for bioremediation of contaminated materials. The Science of Composting, pp. 913–923 (eds. de Bertoldi, M., Sequi, P., Lemmes, B., & Papi, T), Blackie Academic and Professional, London.

33. Clark, G.A.; Stanley, C.D.; Maynard, D.N. (1994). Compost utilization for improved management of vegetable crops on sandy soils-Bradenton site, p. 11–13. In: W.H. Smith (ed.). Summary report for the Florida Composting Conference. Florida Department of Agriculture and Consumer Services, Tallahassee.

34. Coria-Cayupan, Y.S.; De Pinto, M.I.S.; Nazareno, M. A. (2009). Variations in bioactive substance contents and crop yields of lettuce (Lactuca sativa L.) cultivated in soils with different fertilization treatments. Journal of Agricultural and Food Chemistry. 57(21): 10122-10129.

35. Costa, M.S.S.; Sestak, M.; Olibone, D.; Sestak, D.; Kaufmann, A.V.; Rotta, S.R. (2005). Composting of cotton industrial waste. Engenharia-Agricola. 25 (2): 540-548.

36. Domínguez, J. (2004). State of the art and new perspectives on vermicomposting research. In: C.A. Edwards (Ed.). Earthworm Ecology (2nd edition). CRC Press LLC. pp. 401-424.

37. Edwards, C.A.; Lofty, J.R. (1972). Biology of Earthworms. Chapman and Hall, Ltd.: New York.

38. Edwards, C.A.; Burrows, I. (1988). The potential of earthworms composts as plant growth media; In Edward, C.A. and E.F. Neuhauser (Eds.), Earthworms in Waste and Environmental Management. SPB Academic Publishing, The Hague, The Netherlands, ISBN 90-5103-017-7, pp. 21-32.

39. Frelich, L.E.; Hale, C.M.; Sheu, S.; Holdsworth, A.R.; Heneghan, L.; Bohlen, P.J.; Reich, P.B. (2006). Earthworm invasion into previously earthworm - free temperate and boreal forests. Biological Invasions. 8: 1235-1245.

40. Gaddie, R.E.; Douglas, D.E., (1975). Earthworms For Ecology and Profits, Bookworm Publishing Company, Ontario, California, pp. 27-64, 180.

41. Garcia-Gil, J.C.; Plaza, C.; Soler-Rovira, P.; Polo, A. (2000). Long-term effects of municipal solid waste compost application on soil enzyme activities and microbial biomass. Soil Biology and Biochemistry. 32 (13): 1907-1913.

42. Giraddi, R. S. (2000). Influence of vermicomposting methods and season on the biodegradation of organic wastes. Ind. J. of Agric. Sci. 70(10): 663-666.

43. Goenadi, D.H.; Sudharama, I.M. (1995). Shoot initiation by humic acids of selected tropical crops grown in tissue culture. Plant Cell Reports. 15: 59-62.

44. Guerrero, R.D. (2010). Vermicompost production and its use in crop production in the Philippines, Int. J. of Global Environmental Issues ; In Rajiv K. Sinha et. al., (Eds.) Special Issue on 'Vermiculture Technology', 10, 378-383. Inderscience Pub.

45. Gutierrez-Miceli, F.A.; García-Romero, R.C.; Rincon-Rosales, R.; Abud-Archila, M.; Oliva-Llaven, M.A.; Guillen-Cruz, M.J.; Dendooven, L. (2008). Formulation of a liquid fertilizer for sorghum (Sorghum bicolor (L.) Moench) using vermicompost leachate. Bioresource Technology.

46. Gutierrez-Miceli, F.A.; Santiago-Borraz, J.; Montes Molina, J.A.; Nafate, C.C.; Abdud-Archila, M.; Oliva Llaven, M.A.; Rincón-Rosales, R.; Deendoven, L. (2007). Vermicompost as a soil supplement to improve growth, yield and fruit quality of tomato (Lycopersicum esculentum). Bioresource Technology, 98: 2781-2786.

47. Hakkinen, S.H.; Torronen, A.R. (2000). Content of flavonols and selected phenolic acids in strawberries and Vaccinium species: influence of cultivar, cultivation site and technique. Food Res. Int. 33: 517-524.

48. Hale, C.M.; Frelich, L.E.; Reich, P.B. (2005). Exotic European Earthworm Invasion Dynamics in Northern Hardwood Forests of Minnesota, USA. *Ecological Applications*, 15 (3), pp. 848-860.

49. Heaton, S. (2001). Organic farming: food quality and human health; A Review of the Evidence; Soil Assoc. of the United Kingdom, Bristol.

50. Hendrit, P.F., Biological Invasions Belowground : Earthworms as Invasive Species University of Georgia, Athens, Georgia, U.S.A. SPRINGER Reprinted from Biological Invasions, 8 (6): 123, 2006.

51. Jeyabal, A.; Kuppuswamy, G. (2001). Recycling of organic wastes for the production of vermicompost and its response in rice-legume cropping system and soil fertility. European J. Agron. 15(3): 153-170.

52. Joshi, N.V.; Kelkar, B.V. (1952). The role of earthworms in soil fertility. Indian J. of Agricultural Science, 22: 189-196.

53. Kangmin, L.; Peizhen, Li.; Hongtao, Li.(2010). Earthworms helping economy, improving ecology and protecting health. Int. J. of Global Environmental Issues; In Rajiv K. Sinha et. al. (Eds.), Special Issue on 'Vermiculture Technology', 10: 354-365. Inderscience Pub.

54. Lazcano, C.; Gomez-Brandón, M.; Domínguez, J. (2008). Comparison of the effectiveness of composting and vermicomposting for the biological stabilization of cattle manure. Chemosphere. 72: 1013-1019.

55. Lumpkin, H.M. (2005). A comparison of lycopene and other phytochemicals in tomatoes grown under conventional and organic management systems; The World Vegetable Centre, Taiwan; Technical Bull.; AURDC., 34 (4): 48.

56. Magkos, F.; Arvaniti, F.; Zampelas, A. (2006). Organic food: Buying more safety or just peace of mind? A critical review of the literature. Crit. Rev. Food Sci. Nutr. 46: 23-56.

57. Martin, J.P.; Black, J.H.; Hawthorne, R.M. (1999). Earthworm Biology and Production. Circular 455, Florida Cooperative Extension Service, Institute of Food and Agricultural Sciences, University of Florida.

58. Mitchell, A.; Edwards, C. A. (1997). Production of Eisenia fetida and vermicompost from feed-lot cattle manure. Soil Biol. and Biochem. 29(3/4): 763-766.

59. Mitchell, A.E. (2007). Ten-year comparison of the influence of organic and conventional crop management practices on the flavonoids in tomatoes, J. of Agricultural and Food Chemistry.

60. Neilson, R.L. (1965). Presence of plant growth substances in earthworms, demonstrated by the paper chromatography and went pea test. Nature, (London), 208: 1113-1114.

61. Nelson, E.; Rangarajan, A. (2010). Vermicompost: a living soil amendment; Palaniswamy, S. (1996). Earthworm and plant interactions; Paper presented in ICAR Training Program; Tamil Nadu Agricultural University, Coimbatore.

62. Olsson, M.E.; Anderson, C.S.; Oredsson, S.; Berglund, R.H.; Gustavsson, K.E. (2006). Antioxidants levels and inhibition of cancer cells proliferation in-vitro by extracts from organically and conventionally cultivated strawberries, J. of Agricultural Food and Chemistry, 54: 1248 – 1255.

63. OTA. 2006. U.S. organic industry overview. OTA's 2006 manufacturer survey. Organic Trade Assn., Greenfield, Mass.

64. Perucci, P. (1992). Enzymes activity and microbial biomass in field soil amended with municipal refuse. Biol. Fert. Soils; 14: 54-60.

65. Petrussi, F.; de Nobili, M.; Viotto, M.; Sequi, P. (1988). Characterization of organic matter from animal manures after digestion by earthworms. Plant and Soil, 105.

66. Peyvast, G.; Olfati, J.A.; Madeni, S.; Forghani, A. (2008). Effect of vermicompost on the growth and yield of spinach (Spinacia oleracea L.). Journal of Food Agriculture and Environment. 6: 110-113.

67. Pussemier, L.; Larondelle, Y.; Van Peteghem, C.; Huyghebaert, A. (2006). Chemical safety of conventionally and organically produced foodstuffs: a tentative comparison under Belgian conditions. Food Control 17: 14–21.

68. Rajkhowa, D.J.; Gogoi, A.K.;Yaduraju, N.T. (2005). Weed Utilization for Vermicomposting - Success Story National Research Center for Weed Science, Maharajpur, Jabalpur, Madhya Pradesh (India), pp. 4-16.

69. Ramamurthy, V.; Ramesh Kumar, S.C.; Naidu, L.K.G.; Vadivelu, S.; Maji, A.K.; Parhad, V.N. (2007). Agribusiness opportunities in promoting vermicompost in citrus cultivation in Maharashtra. Agric. Econ. Res. Rev. 20: 409-630.

70. Reddy, V.; Katsumi, O. (2004). Vermicomposting of rice straw and its effect on sorghum growth. Tropical Ecology., 45 (2): 327-331.

71. Reddy, B.V.C.; Venkataramanal, M.N.; Kale, R.D.; Balakrishna, A.N. (2006). Impact of application of vermicompost on economics of coconut cultivation in southern Kar. J. Plantation Crops. 34(3): 663-668.

72. Ren, H.; Endo, H.; Hayashi, T. (2001). Antioxidative and antimutagenic activities and polyphenol content of pesticide-free and organically cultivated green vegetables using water-soluble chitosan as a soil modifier and leaf surface spray. J. Sci. Food Agric. 81: 1426-1432.

73. Richard, T.L.; Woodbury, P.B. (1994). What material should be composted? Biocycle, 35 (9), 63.

74. Robbins, M. (2004). Carbon trading, agriculture and poverty; Pub. of World Association of Soil and Water Conservation.,(Special Pub. No. 2): 48 pp.

75. Selden, P.; DuPonte, M.; Sipes, B.; Dinges, K. (2005). Small-Scale Vermicomposting. HG-45.

76. Senesi, N.; Saiz-Jimenez, C.; Miano, T.M. (1992). Spectroscopic characterization of metal-humic acid-like complexes of earthworm-composted organic wastes. The Science of the Total Environment, 117/118, 111-120.

77. Shankar, K.S.; Sumathi, S. (2008). Effect of organic farming on nutritional profile of tomato crops; Central Research Institute for Dryland Agriculture; Hyderabad, India.

78. Sherman, R.; Bambara, S. (1997). Controlling Mite Pests in Earthworm Beds. AGW-001. Raleigh, N.C., Cooperative Extension Service.

79. Sherman, R. (2001). Potential Markets for Vermiculture and Vermicomposting Operations. Vermicomposting News, No. 6.

80. Sherman, R. (1994). Worms Can Recycle Your Garbage. AG-473-18. Raleigh, N.C., Cooperative Extension Service.

81. Sherman-Huntoon, R. (2000). Latest developments in mid-to-large-scale vermicomposting. BioCycle, Vol. 41, No. 11: 51-54.

82. Singh, R.; Gupta, R.K.; Patil, R.T.; Sharma, R.R.; Asrey, R.; Kumar, A.; Jangra, K.K. (2010). Sequential foliar application of vermicompost leachates improves marketable fruit yield and quality of strawberry (Fragaria x ananassa Duch.). Scientia Horticulturae. 124: 34-39.

83. Sinha, R.K.; Agarwal, S.; Chauhan, K.; Valani, D. (2010). The wonders of earthworms & its vermicompost in farm production: Charles Darwin's 'friends of farmers', with potential to replace destructive chemical fertilizers from agriculture. Agricultural Sciences. 1 (2), 76-94.

84. Sinha, R.K.; Hahn, G.; Singh, P.K.; Suhane, R.K.; Anthonyreddy, A. (2011) Organic Farming by Vermiculture: Producing Safe, Nutritive and Protective Foods by Earthworms (Charles Darwin's Friends of Farmers) American Journal of Experimental Agriculture, 1(4): 363-399, 381.

85. Sinha, R.K.; Valani, D.; Chandran, V.; Soni, B.K. (2011). Earthworms - the soil managers: their role in restoration and improvement of soil fertility; Agricultural Issues and Policies; NOVA Science Publishers, Singh, R.; Sharma, R.R.; Kumar, S.; Gupta, R.K.; Patil, R.T. (2008). Vermicompost substitution influences growth, physiological disorders, fruit yield and quality of strawberry (Fragaria x ananassa Duch.). Bioresource Technology. 99: 8507-8511.

86. Slocum, K. (2002). Basic Earthworm Biology. Bon Terra, Inc.: Vancouver, Ore. N.Y., USA; ISBN 978-1-61122-514-3.

87. Stoffella, P.J.; Kahn, B.A., Compost Utilization In Horticultural Cropping Systems, Lewis Publishers Boca Raton London New York Washington, D.C. pp. 133-142.

88. Sunitha, N. D.; Giraddi, R. S.; Lingappa, S.; Megeri, S.N. (1997). Effect of mulches on the activity of earthworm, Eudrilus eugeniae and vermicompost production in black and red soils. Kar. J. Agric. Sci. 10(3): 682-686.

89. Tan, K.H.; D. Tantiwiramanond. (1983). Effect of humic acids on nodulation and dry matter production of soybean, peanut, and clover. Soil Science Society of America Journal. 47: 1121-1124.

90. Tejada, M.; Gonzalez, J.L.; Hernandez, M.T.; Garcia, C. (2008). Agricultural use of leachates obtained from two different vermicomposting processes, Bioresource Technology. 99: 6228-6232.

91. Tomati, V.; Grappelli, A.; Galli, E. (1987). The presence of growth regulators in earthworm worked wastes; In Proc. of Int. Symp. on 'Earthworms'; Italy; 31 March-5 April, 1985., pp. 423-436.

92. Winter, C.K.; Davis, S.F. (2006) Organic Foods, Journal of Food Science, 71 (9), R117-R124.

93. Metzger, J. D.; Shuster, W.; Atiyeh, R. M.; Subler, S.; Edwards, C. A.; Bachman, G., (2000). Effects of vermicomposts and composts on plant growth in horticultural container media and soil, Pedo biologia. 44: 579–590.

94. Mikkonen, T.P.; Maatta, K.R.; Hukkanen, A.T.; Kokko, H.I.; Torronen, A.R.; Karenlampi, S.O.; Karjalainen, R.O. (2001). Flavonol content varies among black currant cultivars. J. Agric. Food Chem. 49: 3274-3277.

95. Nath, G.; Singh, K.; Singh, D.K . (2009). Chemical Analysis of Vermicomposts / Vermiwash of Different Combinations of Animal, Agro and Kitchen Wastes. Australian J. of Basic and Applied Sciences. 3(4): 3672-3676.

96. Nunez-Delicado, E.; Sanchez-Ferrer, A.; Garcia-Carmona, F.F.; Lopez-Nicolas, J.M. (2005). Effect of organic farming practices on the level of latent polyphenol oxidase in grapes. J. Food Sci. 70: C74-C78.

97. Valdrighi, M.M.; Pera, A.; Agnolucci, M.; Frassinetti, S.; Lunardi, D.; Vallini, G. (1996). Effects of compost-derived humic acids on vegetable biomass production and microbial growth within a plant (Cichorium in tybus)-soil system: a comparative study. Agriculture, Ecosystems and Environment. 58: 133-144.

98. Veberic, R.; Trobec, M.; Herbinger, K.; Hofer, M.; Grill, D.; Stampar, F. (2005). Phenolic compounds in some apple (Malus domestica Borkh) cultivars of organic and integrated production. J. Sci. Food Agric. 85: 1687-1694.

99. Vivas, A.; Moreno, B.; Garcia-Rodriguez, S.; Benitez, E. (2009). Assessing the impact of composting and vermicomposting on bacterial community size and structure, and microbial functional diversity of an olive-mill waste. Bioresource Technology. 100: 1319-1326.

100. Wang, D.; Shi, Q.; Wang, X.; Wei, M.; Hu, J.; Liu, J.; Yang, F. (2010). Influence of cow manure vermicompost on the growth, metabolite contents, and antioxidant activities of Chinese cabbage (Brassica campestris ssp. chinensis). Biology and Fertility of Soils. 46: 689-696.

101. Weber, J.; Karczewska, A.; Drozd, J.; Licznar, M.; Licznar, S.; Jamroz, E.; Kocowicz, A. (2007). Agricultural and ecological aspects of a sandy soil as affected by the application of municipal solid waste composts. Soil Biology and Biochemistry. 39: 1294-1302.

102. Young, J.E.; Zhao, X.; Carey, E.E.;Welti, R.; Yang, S-S.; Wang, W. (2005). Phytochemical phenolics in organically grown vegetables. Mol. Nutr. Food Res. 49: 1136-1142.

www.ingramcontent.com/pod-product-compliance
Lightning Source LLC
Chambersburg PA
CBHW021105210326
41598CB00016B/1330